Lecture Notes in Artificial Intelligence 10326

Subseries of Lecture Notes in Computer Science

More information about this series at http://www.springer.com/series/1244

James Cussens · Alessandra Russo (Eds.)

Inductive
Logic Programming

26th International Conference, ILP 2016
London, UK, September 4–6, 2016
Revised Selected Papers

Springer

Editors
James Cussens
Department of Computer Science
University of York
York
UK

Alessandra Russo
Imperial College London
London
UK

ISSN 0302-9743   ISSN 1611-3349 (electronic)
Lecture Notes in Artificial Intelligence
ISBN 978-3-319-63341-1   ISBN 978-3-319-63342-8 (eBook)
DOI 10.1007/978-3-319-63342-8

Library of Congress Control Number: 2017946684

LNCS Sublibrary: SL7 – Artificial Intelligence

Printed on acid-free paper

This Springer imprint is published by Springer Nature
The registered company is Springer International Publishing AG
The registered company address is: Gewerbestrasse 11, 6330 Cham, Switzerland

Preface

This volume constitutes the proceedings of the 26th International Conference on Inductive Logic Programming (ILP 2016) and includes a selection of the papers presented at the conference. ILP 2016 was held in London, during September 4–6, 2016, at the Warren House Conference Centre. Since its first edition in 1991, the annual ILP conference has served as the premier international forum for learning from structured relational data. Originally focusing on the induction of logic programs, over the years it has expanded its research horizon significantly and welcomed contributions on all aspects of learning in logic, multi-relational data mining, statistical relational learning, graph and tree mining, learning in other (non-propositional) logic-based knowledge representation frameworks, exploring intersections with statistical learning, and other probabilistic approaches. Theoretical advances in all these areas have also been accompanied by challenging applications of these techniques to important problems in fields like bioinformatics, medicine, and text mining.

Following the trend of past events, this edition of the conference solicited three types of submissions: (a) long papers describing original mature work containing appropriate experimental evaluation and/or representing a self-contained theoretical contribution; (b) short papers describing original work in progress, brief accounts of original ideas without conclusive evaluation, and other relevant work of potentially high scientific interest but not yet qualifying for the long paper category; and finally (c) papers relevant to the conference topics and recently published or accepted for publication by a first-class conference such as ECML/PKDD, ICML, KDD, ICDM, AAAI, IJCAI, or a journal such as MLJ, DMKD, JMLR etc.

The conference received 35 submissions: ten long papers, 19 short papers, and six published papers. Each of the long and short paper submissions was reviewed by three Program Committee (PC) members. Only four of the ten submitted long papers were accepted for presentation and publication. Short papers were initially evaluated on the basis of the submitted manuscript and the presentation, and authors of a subset of these papers were invited to submit an extended version. After a second review process, only six extended papers were finally accepted for publication. In summary, together with the four long papers, ten papers were accepted to be included in the present volume. The multiple-stage review process, although rather complex, has enabled the selection of high-quality papers for the proceedings. We thank the members of the PC for providing high-quality and timely reviews. Out of all the submitted papers, an additional 13 papers were accepted for publication in the CEUR workshop proceedings series.

The ILP 2016 program included five large technical sessions: Logic and Learning; Graphs and Databases; Probabilistic Logic and Learning; Algorithms, Optimisations and Implementations; and Applications. The papers in this volume represent well the current breadth of ILP research topics such as predicate invention, graph-based learning, spatial learning, logical foundations, statistical relational learning, probabilistic ILP, implementation and scalability, and applications in robotics, cyber-security, and games, providing also an excellent balance across theoretical and practical research. ILP 2016 received generous sponsorship by the *Machine Learning* journal for best student paper awards. The two best student paper awards of ILP 2016 were given to Yi Huang for his paper entitled "Learning Disjunctive Logic Programs from Interpretation Transition," co-authored with Yisong Wang, Ying Zhang and Mingyi Zhang, and to Marcin Malec for his paper "Inductive Logic Programming Meets Relational Databases: An Application to Statistical Relational Learning," co-authored with Tushar Khot, James Nagy, Erik Blasch and Sriraam Natarajan. The conference also received sponsorship from Springer for a best paper award. This award was given to the paper "Generation of Near-Optimal Solutions Using ILP-Guided Sampling" by Ashwin Srinivasan, Gautam Shroff, Lovekesh Vig and Sarmimala Saikia.

With the intent of stimulating collaborations and discussion between academia and industry, the program also featured three invited talks by academic and industrial distinguished researchers. In the talk "Inferring Causal Models of Complex Relational and Dynamic Systems," David Jensen, from the University of Massachusetts, presented key ideas, representations, and algorithms for causal inference, and highlighted new technical frontiers. Frank Wood, from the University of Oxford, gave a talk entitled "Revolutionising Decision Making, Democratising Data Science, and Automating Machine Learning via Probabilistic Programming." In his talk, he gave a broad overview of the emerging field of probabilistic programming, from the point of view of both programming (modelling) language and automated inference, and introduced the most important challenges facing this field. Finally, Vijay Saraswat, senior research scientist in the Cognitive Computing Research division at the IBM T.J. Watson Research Center, discussed in his talk "Machine Learning and Logic: The Beginnings of a New Computer Science?" the open challenges of building cognitive assistants in compliance, and the need to bring together researchers in natural language understanding, machine learning, and knowledge representation/reasoning to address them.

The conference featured, for the first time, an international competition, designed and managed by Mark Law, a member of our local Organizing Committee. The competition was aimed at testing the accuracy, scalability, and versatility of the learning systems that were entered. The competition had two main tracks for probabilistic and non-probabilistic approaches. The winners of the competition were Peter Schüller, from Marmara University, for his non-probabilistic approach and jointly Riccardo Zese, Elena Bellodi, and Fabrizio Riguzzi for their probabilistic approach. Results of the competition are publicly available on http://ilp16.doc.ic.ac.uk/competition.

The ILP 2016 conference was kindly sponsored by IBM Watson Research, the Association of Logic Programming, Springer's *Lecture Notes in Artificial Intelligence*, the *Artificial Intelligence* journal, and the *Machine Learning* journal. We would like to thank EasyChair for supporting the submission handling. We would like to thank the members of the local Organizing Committee of ILP 2016: Krysia Broda, Dalal Alrajeh, and Mark Law. Our thanks also go to Mark Law for running the competition and for setting up and maintaining the website. The conference would not have been possible without their hard work.

Finally, we would like to thank all those involved in making ILP 2016 such a success: our invited speakers, our sponsors, the PC and, of course, those who came to ILP 2016 to present and discuss their work.

May 2017   James Cussens
                                                                              Alessandra Russo

Organization

Organizing Committee

Program Co-chairs

James Cussens   University of York, UK
Alessandra Russo   Imperial College, UK

Competition Chair

Mark Law   Imperial College, UK

Financial Chair

Dalal Alrajeh   Imperial College, UK

Publicity Chair

Krysia Broda   Imperial College, UK

Program Committee

Dalal Alrajeh	Imperial College London, UK
Alexander Artikis	Institute of Informatics and Telecommunications, Greece
Krysia Broda	Imperial College London, UK
Rui Camacho	LIACC/FEUP University of Porto, Portugal
Luc De Raedt	Katholieke Universiteit Leuven, Belgium
Sašo Džeroski	University of Ljubljana, Slovenia
Floriana Esposito	Università degli Studi di Bari, Italy
Nicola Fanizzi	Università degli Studi di Bari, Italy
Stefano Ferilli	Università degli Studi di Bari, Italy
Nuno Fonseca	European Bioinformatics Institute, Portugal
Katsumi Inoue	National Institute of Informatics, Japan
Kristian Kersting	TU Dortmund University, Germany
Ross King	University of Manchester, UK
Nicolas Lachiche	University of Strasbourg, France
Nada Lavrač	Jožef Stefan Institute, Slovenia
Francesca Lisi	Università degli Studi di Bari, Italy
Donato Malerba	Università degli Studi di Bari, Italy
Stephen Muggleton	Imperial College London, UK
Aline Paes	UFF, Federal Fluminense University, Brazil
Jan Ramon	Inria, France
Oliver Ray	University of Bristol, UK

Fabrizio Riguzzi	University of Ferrara, Italy
Chiaki Sakama	Wakayama University, Japan
Vítor Santos Costa	Universidade do Porto, Portugal
Takayoshi Shoudai	Kyushu International University, Japan
Alireza Tamaddoni-Nezhad	Imperial College London, UK
Christel Vrain	University of Orleans, France
Stefan Wrobel	Fraunhofer IAIS and University of Bonn, Germany
Akihiro Yamamoto	Kyoto University, Japan
Gerson Zaverucha	PESC-COPPE, UFRJ, Brazil
Filip Železný	Czech Technical University, Czech Republic

Sponsoring Institutions

Association for Logic Programming
IBM Research
Machine Learning Journal (for best student paper awards)
Lecture Notes in Artificial Intelligence, Springer (for best paper award)
Artificial Intelligence Journal

Invited Speakers

Inferring Causal Models of Complex Relational and Dynamic Systems

David Jensen

Knowledge Discovery Laboratory, Computational Social Science Institute,
College of Information and Computer Sciences,
University of Massachusetts Amherst, Amherst, USA

Over the past 25 years, surprisingly effective techniques have been developed for inferring causal models from observational data. While traditional models reason about a given system by assuming that its behavior is stationary, causal models reason about how a system will behave under intervention. Unfortunately, nearly all existing methods for causal inference assume that data instances are independent and identically distributed, making them inappropriate for analyzing many social, economic, biological, and computational systems. In this talk, I will explain the key ideas, representations, and algorithms for causal inference, and I will describe very recent developments that extend those techniques to complicated systems with relational and dynamic behavior. I will describe practical methods for evaluating methods for causal inference and identify some of the most pressing research questions and new technical frontiers.

Machine Learning and Logic—The Beginnings of a New Computer Science?

Vijay A. Saraswat

IBM T.J. Watson Research Lab, New York, USA

Our long-term research goal in Cognitive Computing Research at IBM is to develop systems that know deeply, learn continuously, reason with purpose and interact naturally. To further this agenda, we are focusing on a few deep domains. This talk will address the challenges of building cognitive assistants in compliance assistants that deal with understanding and reasoning about the myriad (corporate, financial, privacy, ethical) laws and regulations within the context of which modern international businesses must operate. An interim goal for the compliance cognitive assistant is to clear the Uniform CPA exam, a professional certification attempted by master's level students. We will outline the tremendous technical challenges underlying this goal and our current approaches. We believe the key to achieving this goal is bringing together researchers in natural language understanding, machine learning, and knowledge representation/reasoning for a concerted attack on this problem.

Revolutionizing Decision Making, Democratizing Data Science, and Automating Machine Learning via Probabilistic Programming

Frank Wood

Department of Engineering Science, University of Oxford, Oxford, UK

Probabilistic programming aims to enable the next generation of data scientists to easily and efficiently create the kinds of probabilistic models needed to inform decisions and accelerate scientific discovery in the realm of big data and big models. Model creation and the learning of probabilistic models from data are key problems in data science. Probabilistic models are used for forecasting, filling in missing data, outlier detection, cleanup, classification, and scientific understanding of data in every academic field and every industrial sector. While much work in probabilistic modeling has been based on hand-built models and laboriously-derived inference methods, future advances in model-based data science will require the development of much more powerful automated tools than currently exist. In the absence of such automated tools, probabilistic models have traditionally co-evolved with methods for performing inference. In both academic and industrial practice, specific modeling assumptions are made not because they are appropriate to the application domain, but because they are required to leverage existing software packages or inference methods. This intertwined nature of modeling and computation leaves much of the promise of probabilistic modeling out of reach for even expert data scientists. The emerging field of probabilistic programming will reduce the technical and cognitive overhead associated with writing and designing novel probabilistic models by both introducing a programming (modeling) language abstraction barrier and automating inference. The automation of inference, in particular, will lead to massive productivity gains for data scientists, much akin to how high-level programming languages and advances in compiler technology have transformed software developer productivity. What is more, not only will traditional data science be accelerated, but the number and kind of people who can do data science also will be dramatically increased. My talk will touch on all of this, explain how to develop such probabilistic programming languages, highlight some exciting ways such languages are starting to be used, and introduce what I think are some of the most important challenges facing the field as we go forward.

Contents

Estimation-Based Search Space Traversal in PILP Environments. 1
    Joana Côrte-Real, Inês Dutra, and Ricardo Rocha

Inductive Logic Programming Meets Relational Databases: Efficient
Learning of Markov Logic Networks. 14
    Marcin Malec, Tushar Khot, James Nagy, Erik Blask,
    and Sriraam Natarajan

Online Structure Learning for Traffic Management 27
    Evangelos Michelioudakis, Alexander Artikis, and Georgios Paliouras

Learning Through Advice-Seeking via Transfer 40
    Phillip Odom, Raksha Kumaraswamy, Kristian Kersting,
    and Sriraam Natarajan

How Does Predicate Invention Affect Human Comprehensibility?. 52
    Ute Schmid, Christina Zeller, Tarek Besold, Alireza Tamaddoni-Nezhad,
    and Stephen Muggleton

Distributional Learning of Regular Formal Graph System
of Bounded Degree. 68
    Takayoshi Shoudai, Satoshi Matsumoto, and Yusuke Suzuki

Learning Relational Dependency Networks for Relation Extraction 81
    Ameet Soni, Dileep Viswanathan, Jude Shavlik, and Sriraam Natarajan

Towards Nonmonotonic Relational Learning from Knowledge Graphs. 94
    Hai Dang Tran, Daria Stepanova, Mohamed H. Gad-Elrab,
    Francesca A. Lisi, and Gerhard Weikum

Learning Predictive Categories Using Lifted Relational Neural Networks. . . . 108
    Gustav Šourek, Suresh Manandhar, Filip Železný, Steven Schockaert,
    and Ondřej Kuželka

Generation of Near-Optimal Solutions Using ILP-Guided Sampling. 120
    Ashwin Srinivasan, Gautam Shroff, Lovekesh Vig, and Sarmimala Saikia

Author Index 133

Estimation-Based Search Space Traversal in PILP Environments

Joana Côrte-Real[✉], Inês Dutra, and Ricardo Rocha

Faculty of Sciences, CRACS, INESC TEC, University of Porto,
Rua do Campo Alegre, 1021/1055, 4169-007 Porto, Portugal
{jcr,ines,ricroc}@dcc.fc.up.pt

Abstract. Probabilistic Inductive Logic Programming (PILP) systems extend ILP by allowing the world to be represented using probabilistic facts and rules, and by learning probabilistic theories that can be used to make predictions. However, such systems can be inefficient both due to the large search space inherited from the ILP algorithm and to the probabilistic evaluation needed whenever a new candidate theory is generated. To address the latter issue, this work introduces probability estimators aimed at improving the efficiency of PILP systems. An estimator can avoid the computational cost of probabilistic theory evaluation by providing an estimate of the value of the combination of two subtheories. Experiments are performed on three real-world datasets of different areas (biology, medical and web-based) and show that, by reducing the number of theories to be evaluated, the estimators can significantly shorten the execution time without losing probabilistic accuracy.

1 Introduction

Probabilistic Inductive Logic Programming (PILP) [4] is an extension of the ILP paradigm that can represent knowledge using probabilistic facts and rules and which learns, as a result, probabilistic theories that can be used for prediction. Introducing probabilistic information in ILP to create PILP can be used to (i) create better logical models that can take uncertainty into account; (ii) implicitly reduce the theory search space by transforming numerical arguments in annotated probabilistic data; (iii) compress data by representing it as aggregates; or (iv) add knowledge from the literature in the form of probabilistic information. PILP can be seen as a Statistical Relational Learning approach (SRL) [4] and, in this setting, both parameter and structure learning are possible. However, it is more common for SRL techniques to learn parameters, and only few SRL methods can learn structure, or both. PILP differs from other SRL techniques because it focuses primarily on structure learning over relational data that is already annotated with probabilistic values.

PILP suffers from the same search space traversal efficiency issues as ILP because similar algorithms are used to generate the logical part of the theories. Additionally, PILP adds a level of complexity because every new theory generated needs to be probabilistically evaluated in order to be considered. This

© Springer International Publishing AG 2017
J. Cussens and A. Russo (Eds.): ILP 2016, LNAI 10326, pp. 1–13, 2017.
DOI: 10.1007/978-3-319-63342-8_1

work presents a strategy aimed at improving the performance of PILP systems through the use of *estimators* that can prune the universe of candidate theories and, thus, reduce the search space. These estimators were integrated in the SkILL system [2], but the concepts are general to any PILP engine.

SkILL is a stochastic inductive logic learner which can generate First-Order Logic (FOL) theories based on a database of probabilistic data. These theories are expressed as Horn clauses (a subset of FOL) and so they can be used to extract relational non-trivial knowledge about the dataset where they are inferred from. SkILL differs from other PILP systems such as ProbFOIL+ [3] or SLIPCOVER [1] because it introduces an algorithm of polynomially bound complexity on user-defined parameters, as well as a number of efficient pruning strategies that can reduce execution time while maintaining prediction quality.

To the best of the authors' knowledge, the notion of an estimator is a novel feature in PILP systems. In this work, five estimators that can be incorporated in the estimation pruning strategy are proposed, namely *minimum*, *maximum*, *center*, *independence* and *exclusion*. To validate this estimation-based search space traversal approach, a thorough experimental analysis of the impact that each estimator has on the execution time and theory quality is presented. Experiments are performed in three probabilistic datasets, and the models are validated using hold-out resampling or leave-one-out cross-validation techniques. Results show that estimators can significantly prune the search space, and thus, reduce execution time, while maintaining the same probabilistic accuracy when compared with using no estimation pruning.

2   Related Work

According to Getoor *et al.* [8], relational data introduces the machine learning problem of *class-level frequency estimation*: building a model that can answer generic statistical queries about classes of individuals in a database. This is opposed to *instance-level frequency estimation*, where one is interested in the probability of a particular instance. In a first-order logic representation, the first type of estimation would be described with first-order formulas with variables, while the second type would be described with first-order formulas with constants (ground terms). There has been a whole body of research on modeling relational data using various kinds of representations, inference systems, and learning techniques (parameters and structure) [7].

There are many *probabilistic languages* that can represent and perform inference with probabilities, such as SLP [12], BLP [9], CLP($\mathcal{BN}$) [15], ProbLog [10], MLN [14], Prism [16], among others (for a recent survey of probabilistic logic languages see [5]). Probabilistic logic languages have been around for over 20 years [5]. They differ in the way they extend logic to include probabilities (class-level or instance-level), in their syntax, in the kind of uncertainty that is represented (probabilities, weights or potential functions), and in their inference algorithms. However, there are few works in the probabilistic logic field dedicated to learning the structure of classifiers using a human-readable probabilistic representation for knowledge.

An example of a system that uses logic to learn from probabilistic representations is MLN, where the learning algorithm is ILP-based and uncertainty is represented as potential functions [11]. Several methods of inference can be used and the models learnt are first-order logic formulas with potential scores. Another ILP-based example is Natarajan *et al.*'s boosting approach [13], which uses regression trees to learn the model structure faster. Furthermore, some works in the literature allow the representation of probabilistic logic using Bayesian networks. This is the case of Schulte *et al.*'s PBN (Parametrized Bayesian Networks) [17]. PBN is a first-order logic extension of Bayesian networks, where nodes are represented as a first-order term with a variable.

This work's focus is on *probabilistic inductive logic programming* (PILP) systems. These systems use as basis a probabilistic logic language to learn (probabilistic) theories. This work follows the syntax and inference mechanism of ProbLog [10], and uses it to represent the datasets and to learn first-order (probabilistic) theories. ProbLog is an extension of Prolog, whose syntax is modified to take into account class-level and instance-level probabilities, and annotated disjunctions. Uncertainty is, thus, represented as probabilities.

There are several PILP approaches mentioned in the literature, such as Prob-FOIL/ProbFOIL+ [3], SLIPCOVER [1] and SkILL [2]. ProbFOIL+ is a PILP system that can tune the prediction of a theory by finding a weight for each rule in that theory. ProbFOIL+ algorithm computes the best weight whenever a rule is being added to the theory and then integrates it in the theory. This can be seen as a form of *boosting*, since the importance of each rule in the theory is being adjusted, even though the possibility for adjustment is limited (the weight must be between 0 and 1). SLIPCOVER introduces the new ability to perform generative learning in the search space. SLIPCOVER still requires a target predicate, but it also gathers a set of good theories which can explain predicates from the BK other than the target predicate – this process can be viewed as a form of deep learning, since these intermediate theories will be used to explain the target predicate. SkILL is a PILP system which introduces an algorithm of polynomially bound complexity on user-defined parameters, as well as a number of efficient pruning strategies that can reduce execution time while maintaining prediction quality. A comparison between SkILL, ProbFOIL+ and an ILP system can be found in [2].

3   Background

PILP extends the ILP setting by introducing Probabilistic Background Knowledge (PBK), where FOL data descriptions can be annotated with a probability value ranging from 0 to 1, and Probabilistic Examples (PE), no longer positive or negative, also with a value ranging between 0 and 1. Because PILP theories are still generated based on the logical information of the data, the ILP language bias translates directly to PILP. The process of generating theories also mimics ILP, since they are based on the logical clauses in the PBK. Therefore the search space algorithm of PILP has the same efficiency issues of ILP's. Furthermore,

PILP adds an extra level of complexity due to the probabilistic evaluation of theories w.r.t. the examples. The background knowledge can be composed of (Horn) clauses, which can be facts or definite clauses Definite clauses are composed of a head and a body, and the body represents the explanation for the head. Facts and definite clauses' heads are examples of *literals* that ILP and PILP use to build rules.

As mentioned before, in this work, probabilities are annotated according to ProbLog's syntax [10]. Each clause $p_j :: c_j$ in the PBK represents an independent binary random variable in ProbLog, meaning that it can either be true with probability p_j or false with probability $1 - p_j$. Each set of possible choices over all clauses of the PBK represents a *possible world* ω_i, where ω_i^+ is the set of clauses that are true in that particular world, and $\omega_i^- = \omega_i \backslash \omega_i^+$ is the set of clauses that are false. Since these clauses have a probabilistic value, a ProbLog program defining a probabilistic distribution over the possible worlds can be formalized as shown in Eq. 1. A ProbLog *query* q is said to be true in all worlds w^q where $w^q \models q$, and false in all other worlds. As such, the *success probability* of a query is given by the sum of the probabilities of all worlds where it is found to be true, as denoted in Eq. 2.

$$P(\omega_i) = \prod_{c_j \in \omega_i^+} p_j \prod_{c_j \in \omega_i^-} (1 - p_j) \tag{1}$$

$$P(q) = \sum_{\omega_i \models q} P(\omega_i) \tag{2}$$

One important difference between ILP and PILP lies in the assessment of the fitness of theories – in PILP the *loss function* must be able to evaluate probabilistic inputs. As such, the aim of PILP systems is to find theories which most closely predict the value of the examples (also ranging between 0 and 1), or rather that minimize the error between predictions and the examples' values.

Theories can be formed either by a single rule (clause) or by a set of rules (where the clauses are mutually disjunctive). The length of a theory is equal to the number of rules it contains. In SkILL, theories can be combined using either the AND or the OR operation, which correspond to the logical conjunction and disjunction of the rules in the theories, respectively. In the case of the AND operation, only single rules (theories of length one) can be combined, and the result is another theory of length one (e.g. combining theories $t(X):- p(X)$ and $t(X):- q(X, Y)$ using the AND operation would result in theory $t(X):- p(X), q(X,Y))$[1]. Conversely, theories of any length can be combined using the OR operation, and the resulting theory's length is equal to the sum of the lengths of the combined theories (e.g. combining theories of length one $t(X):- r(X)$ and $t(X):- s(X, Y)$ using the OR operation would result in theory $t(X):- r(X); s(X, Y)$ of length 2).

---

[1] The unification of variables between the literals is obtained from the specified language bias.

SkILL's algorithm is composed of two main steps: (i) building theories of length one (single rules) using the AND operation, and (ii) building theories of length greater than one using the OR operation. In step (i), single rules of increasing number of literals are built from the mode declarations using the AND operation. Adding literals to a rule in conjunction makes the resulting rule more specific. Once all possible rules are built and evaluated, the algorithm proceeds to step (ii) using the OR operation to combine single rules (theories of length one) into theories of greater length, up to a maximum length. By adding rules to a theory in disjunction, the resulting theory becomes more general.

In order to assess a theory's fitness, its exact probabilistic value for each example must be computed, so that the theory is *evaluated exactly*. This process can be very time consuming, since the evaluation process must consider all possible worlds where the theory may be true. For a small number of facts in the PBK this is not a problem, but exact computation grows exponentially as the size of the PBK is increased. Consider the process of evaluating exactly the theory $t(X):- p(X), q(X,Y)$. ProbLog would need to compute all possible worlds for this theory in order to assess the overall error of the theory's predictions against the examples. Whether the theory is stored for further combinations or discarded after the evaluation stage, the system has already spent a considerable amount of time just to evaluate it.

To mitigate this problem, this work introduces the *estimation pruning* strategy, which can discard theories based on their previously evaluated subparts. For instance, suppose that theories $t(X):- p(X)$ and $t(X):- q(X,Y)$ had already been evaluated – in that case, it is possible to make an estimation of the value of $t(X):- p(X), q(X,Y)$ based on this information. Thus, estimation pruning consists of ruling out theories that have poor estimations and exactly evaluating theories that have good estimations. In SkILL, the decision on whether a theory is discarded is made based on one of two criteria: *soft pruning* or *hard pruning*. After the initial step of estimating the values for each example, the estimated value's usefulness is assessed according to one of these criteria. Note that the criteria are directly applicable to the estimated probabilistic values in lieu of the exact predictions of a theory. The combination is then pruned away if it is found to be useless. Conversely, if the combination is considered useful, then exact probabilistic evaluation is performed and the theory and its exact evaluation are saved for the next iteration.

4   Estimation Pruning

Estimation pruning consists of estimating the predictions of two theories combined based on the individual predictions of each theory. Estimation pruning excludes combinations of theories whose *estimated predictions* suggest that the resulting theory will be too specific (for the AND operation) or too general (for the OR operation). This process is somehow similar to the evaluation of theories in ILP. For instance, the more specific theory t_s will not cover more positive examples than a more general theory t_g and so it can be discarded.

In the PILP setting, the exact probabilistic evaluation of a theory corresponds to the weighted proportion of worlds where the theory is true. The probabilistic value for an example e using a theory t is given by determining in how many worlds (of all possible worlds in the PBK) $t(e)$ is true. The challenge in estimating the value of a probabilistic evaluation knowing the values of the theories being combined lies in the fact that the *amount of overlapping* of the sets of worlds corresponding to those two theories is unknown before evaluation. If two theories are mutually exclusive (or disjoint) w.r.t. the PBK, then their overlap is null. On the other hand, if a theory is more specific than another, the former will cover a subset of the worlds covered by the latter. Theories can also be independent, meaning that the probability that one theory is true in a world does not change the probability that another theory is also true in that world.

Despite this uncertainty, it is possible to calculate the interval where the predictions of a combination of two theories will be (this is depicted in Fig. 1 as a shaded area). The lower and upper bounds of the interval are determined by the predictions of the theories that are being combined (t_1 and t_2 in Fig. 1[2]). Depending on where the resulting theory t will lie in the interval, the (vertical) distance between t's values and the example values (squares in Fig. 1) will vary, and as t converges to the examples, its prediction quality is improved.

This work presents five estimators that can be used to estimate the value of theories, namely: *minimum, maximum, center, independence* and *exclusion*. These estimators predict different sets of values inside the estimation interval, based on different set theory cases. The *minimum* and *maximum* estimators correspond to the lower and upper boundaries of the estimation interval (*min* and *max* estimators in Fig. 1, respectively). The *center* estimator (*ctr* in Fig. 1) is the center of the estimation interval (halfway between *minimum* and *maximum*). The *independence* estimator (*ind* in Fig. 1) assumes that theories t_1 and t_2 are independent and calculates the values of their combination accordingly. The *exclusion* estimator (not depicted in Fig. 1) assumes that the theories t_1 and t_2 are as exclusive as possible. In the AND operation, the *exclusion* estimator is equal to the *minimum* estimator, since when two theories are mutually exclusive, their amount of overlap is minimum. The first row in Table 1 summarizes the expressions used to calculate these estimations.

Table 1. Expressions used to calculate estimations

Operation	Minimum	Maximum	Center	Independence	Exclusion
AND	$max(0, A + B - 1)$	$min(A, B)$	$\frac{1}{2}(min(A, B) + max(0, A + B - 1))$	$A \times B$	$max(0, A + B - 1)$
OR	$max(A, B)$	$min(A + B, 1)$	$\frac{1}{2}(max(A, B) + min(A + B, 1)$	$A + B - A \times B$	$min(A + B, 1)$

---

[2] Theories are indexed only for clarity's sake. They correspond to the same concept to be learned.

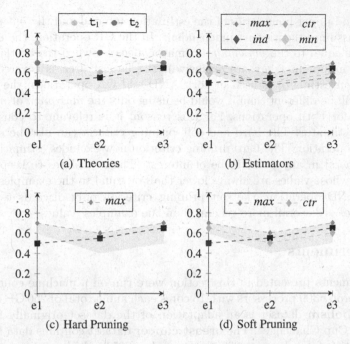

Fig. 1. Estimators in AND operation. The x-axis contains three examples and the y-axis represents probabilistic values. Examples are depicted as squares, theories t_1 and t_2 as circles and estimators *min, max, ctr, ind* as diamonds.

After calculating an estimation for the combination of theories t_1 and t_2, it is necessary to decide, based on the estimation, whether the combination of theories should be evaluated. Thus, two pruning criteria can be used: hard pruning (Fig. 1(c)) or soft pruning (Fig. 1(d)). For the AND operation, the hard pruning criterion discards theories that are too specific in any of their predictions. This means that the estimations must be higher than the examples' values in every point (in Fig. 1(c) this only happens if the *maximum* estimator is being used to estimate the combination). The soft pruning criterion only prunes the theory away if it is *overall* more specific than the example values. In Fig. 1(d), the estimators that are not discarded are those that are above (*maximum*) or equally above and below (*center*) the examples' values. Estimator *center* is kept because its estimations are just a small distance below two example values but are a large distance above the first example value, which balances out. Pruning combinations of theories can be extended to the OR operation. Like the AND operation, this strategy estimates the value of a combination of two theories. In the OR setting, theories are excluded when they are found to be too general to benefit from further combination. Based on the expressions presented in Table 1 and following a similar reasoning to the AND operation, the same five estimators can be defined. Again, the *minimum* and *maximum* estimators define the estimation interval based on t_1 and t_2. The *center* estimator is the value halfway between the lower and upper boundaries of the estimation interval and the *independence* estimator assumes theories are independent. In the OR operation, the *exclusion* estimator is equal to the *maximum* estimator, because when the overlap of two theories is minimum (they are exclusive), the largest area is covered. When the *exclusion* estimator is used in both AND and OR operations, the result it produces will be different than it would be using only the *maximum* or *minimum* estimators for both operations. For this reason, it is relevant to consider the *exclusion* estimator. The hard and soft pruning criteria can also be extended to the OR operation. The hard pruning criterion now excludes estimations that are too general in any point to be of interest. This translates to keeping only estimators whose values are always lower than or equal to the examples' values. Like the AND operation, the soft pruning criterion only discards estimators whose values are overall more general than the examples' values.

5   Experiments

The experiments presented in this section were run on a machine containing 4 AMD Opteron 6300 processors with 16 cores each and a total of 250 GB of RAM. The **metabolism** dataset is an adaptation of the dataset originally from the 2001 KDD Cup Challenge[3]. The **breast cancer** dataset contains data from 130 biopsies dating from January 2006 to December 2011, which were prospectively given a non-definitive diagnosis at radiologic-histologic correlation conferences. The **athletes** dataset consists of a subset of facts regarding athletes and the sports they play collected by the never-ending language learner NELL[4]. For the **metabolism** and **athletes** datasets, a number of n-times hold-out sets were made and all measurements were averaged out over the folds. In the **breast cancer** dataset, leave-one-out cross-validation was used.

Different combinations of estimation pruning were tested: only pruning the AND operation, only pruning the OR operation, and pruning both operations. The pruning settings are reported as a set of two letters: the first letter is the AND pruning option and the second is the OR pruning option. Pruning options can be soft pruning (S), hard pruning (H) or no pruning (x). For example, using this codification, xS stands for no AND pruning and soft OR pruning. For each configuration, several measurements were recorded for each dataset: execution time, probabilistic accuracy on the test set, and number of rules and theories pruned. The probabilistic accuracy metric used in this work is equivalent to the mean absolute error of predictions calculated against example values, and was first introduced by De Raedt and Thon in [6].

Tables 2, 3 and 4 present the speedups and ratio of probabilistic accuracy for the **metabolism, breast cancer** and **athletes** datasets, respectively. The speedup $\frac{B_t}{P_t}$ is calculated w.r.t. the B_t base case time (no pruning) for different P_t pruning options' execution time. If there is a slowdown, the inverse speedup $\frac{P_t}{B_t}$ is presented as a negative number. The ratio of the probabilistic accuracy $\frac{P_a}{B_a}$

---

[3] http://www.cs.wisc.edu/~dpage/kddcup2001.
[4] http://rtw.ml.cmu.edu.

Table 2. Speedup and probabilistic accuracy ratio for **metabolism** dataset

Speedup

Est	Sx	Hx	xS	xH	SS	HH
min	1.45	1.47	−1.03	1.36	1.47	2.52
max	1.56	1.56	−1.09	1.94	1.61	5.66
ctr	1.57	1.58	1.03	1.95	1.61	5.65
ind	1.35	1.33	−1.23	1.64	1.46	5.37
exc	1.57	1.58	−1.04	1.95	1.58	5.69

Probabilistic accuracy ratio

Est	Sx	Hx	xS	xH	SS	HH
min	1.00	1.00	1.00	1.01	1.00	1.00
max	1.00	1.00	1.00	1.00	1.00	−1.01
ctr	1.00	1.00	1.00	1.00	−1.01	−1.01
ind	1.00	1.00	1.00	1.00	1.00	−1.01
exc	1.00	1.00	1.00	1.00	1.00	−1.01

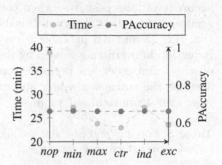

Fig. 2. Average time (in minutes) and probabilistic accuracy in **metabolism** dataset, for the base case (*nop*) and the five estimators. Values for each estimator are the average of its result over the pruning options.

is calculated for each probabilistic settings P_a w.r.t. the probabilistic accuracy of the B_a base case. Similarly to the speedup, when the probabilistic accuracy decreases, the inverse of the ratio is given $\frac{B_a}{P_a}$ as a negative number. Figures 2, 3 and 4 depict the variation in execution time in minutes (left y-axis) and the variation in probabilistic accuracy (right y-axis) for all estimators in the **metabolism**, **breast cancer** and **athletes** datasets, respectively. The estimators analysed were the base case (no estimation pruning performed, or *nop*), *minimum* (*min*), *maximum* (*max*), *center* (*ctr*), *independence* (*ind*), and *exclusion* (*exc*). Each dataset's results will be discussed next.

For the **metabolism** dataset, results in Table 2 show that the greatest reduction in execution time is achieved by all estimators in the HH pruning setting. The xS pruning setting shows the slowest execution times with all estimators, except *center*, causing a slowdown. There is no significant reduction in probabilistic accuracy in any setting. Figure 2 shows that, overall, the probabilistic accuracy of the theories is unchanged and that the *maximum*, *center* and *exclusion* estimators can all reduce execution time from 40 to less than 25 min.

Results in Table 3 show that, in the **breast cancer** dataset, the greatest reduction in execution time can be achieved by using pruning in both the AND and the OR operations (SS and HH settings). The pruning settings that use only OR pruning (xS and xH) present more modest reductions of execution time (about 1.5 times) when compared to the settings that use only AND pruning (about 7 times). Although the OR operation has the potential to increase the (probabilistic) accuracy of true positives, it may also increase the accuracy of false positives. On the other hand, the AND operation, for this domain, works better, since it maintains the accuracy of true positives while decreasing the accuracy of false positives, when combining literals in a theory. The predictive accuracy of the best theory in this dataset never decreases, and in some settings (Sx, Hx, SS and HH in Table 3) even increases slightly. This effect is due to a reduction in overfitting caused by the exclusion of some theories that are better on the training set but perform worse on the test set. Figure 3 shows that, on average, the *maximum*, *center* and *exclusion* datasets can reduce execution time from over 4 min to about 1 min.

Table 3. Speedup and probabilistic accuracy ratio for **breast cancer** dataset

Speedup						
Est	Sx	Hx	xS	xH	SS	HH
min	7.09	7.18	1.42	1.41	22.44	21.46
max	7.16	7.06	1.65	1.63	25.24	23.49
ctr	7.04	6.98	1.63	1.63	25.25	24.80
ind	7.20	7.02	1.42	1.29	22.62	22.84
exc	7.19	7.19	1.63	1.62	25.00	24.80
Probabilistic accuracy ratio						
Est	Sx	Hx	xS	xH	SS	HH
min	1.09	1.09	1.00	1.00	1.09	1.09
max	1.09	1.09	1.00	1.00	1.09	1.09
ctr	1.09	1.09	1.00	1.00	1.09	1.09
ind	1.09	1.00	1.00	1.00	1.09	1.09
exc	1.09	1.09	1.00	1.00	1.09	1.09

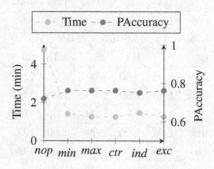

Fig. 3. Average time (in minutes) and probabilistic accuracy in **breast cancer** dataset, for the base case (*nop*) and the five estimators. Values for each estimator are the average of its result over the pruning options.

Table 4. Speedup and probabilistic accuracy ratio for **athletes** dataset

Speedup						
Est	Sx	Hx	xS	xH	SS	HH
min	3.34	3.33	1.01	1.66	3.63	9.48
max	3.35	3.35	2.12	2.19	12.40	49.72
ctr	3.20	3.28	1.00	1.80	3.62	19.82
ind	3.33	3.34	2.10	2.17	12.34	50.36
exc	3.31	3.23	2.02	2.11	11.86	48.01
Probabilistic accuracy ratio						
Est	Sx	Hx	xS	xH	SS	HH
min	−1.06	−1.06	1.00	1.00	−1.11	1.00
max	−1.06	−1.06	1.00	1.00	−1.15	1.00
ctr	−1.06	−1.06	1.00	1.00	−1.08	1.00
ind	−1.06	−1.06	1.00	1.00	−1.15	1.00
exc	−1.06	−1.06	1.00	1.00	−1.15	1.00

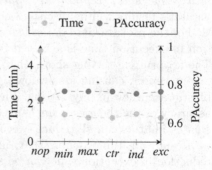

Fig. 4. Average time (in minutes) and probabilistic accuracy in **athletes** dataset, for the base case (*nop*) and the five estimators. Values for each estimator are the average of its result over the pruning options.

In the **athletes** dataset, again the HH pruning setting can reduce most execution time. However, the reduction using estimators *minimum* and *center* is much less than that of estimators *maximum, independence* and *exclusion*, where the execution is about 50 times faster (Table 4). Estimators *minimum* and *center* are consistently slower in other pruning settings (xS, xH and SS), and the xS and xH settings present the lowest reduction in execution time in this dataset, of 2 times on average. Similarly to the other datasets, Table 4 shows that the probabilistic accuracy in the **athletes** dataset presents no significant reduction and, in particular, in the xS, xH and HH settings it is not reduced at all. Estimators *maximum, independence* and *exclusion* present the greatest overall reduction in execution time (Fig. 4), from 20 to about 5 min, on average.

Finally, for the **athletes** dataset, Table 5 presents the number of probabilistic evaluations performed for each pruning setting and estimator. The first number corresponds to single rules (theories of length one) evaluated, and thus the reduction is caused by AND pruning. Similarly, the second number in each cell is the number of theories of length greater than one, and its reduction is caused by OR pruning. The greatest reductions correspond to the HH setting (column 8 in Table 5), and are consistent with the settings in Table 4 that presents the greatest speedups. Additionally, from Fig. 4, the three fastest estimators (in average) are also the estimators that in Table 5 prune away most theories. In particular, the number of theories pruned away during OR pruning is significantly lower for estimators *max, ind* and *exc* when compared to estimators *min* and *ctr*. The same trend can be observed in the other datasets but results were omitted due to lack of space.

Table 5. Number of single rules/theories evaluated for the **athletes** dataset

Est	xx	Sx	Hx	xS	xH	SS	HH
min	2414/1989	164/968	164/968	2414/1981	2414/604	164/913	164/361
max	2414/1989	164/968	164/968	2414/69	2414/0	164/243	164/0
ctr	2414/1989	164/968	164/968	2414/1974	2414/381	164/907	164/128
ind	2414/1989	164/968	164/968	2414/69	2414/0	164/243	164/0
exc	2414/1989	164/968	164/968	2414/69	2414/0	164/243	164/0

6   Conclusion

This work proposed five PILP estimators whose aim is to alleviate the overhead imposed by the exact evaluation of combinations of candidate probabilistic theories. Because PILP theories can be built using both conjunction (AND operation) and disjunction (OR operation), the estimators must be adapted accordingly. The estimators were implemented in the estimation pruning stage of the SkILL system, but can be generalized to any PILP engine. Results showed that all estimators resulted in faster execution times when coupled with an H (hard) pruning setting and, in particular, the HH pruning setting showed the greatest speedups and also the greatest reduction in the number of probabilistic evaluations performed. Even though all estimators maintain predictive quality and reduce execution time, estimators *maximum* and *exclusion* are overall faster, and opting for one of these estimators in lieu of estimators *minimum, center* or *independence* can result in an up to 5 times faster runtime for the same pruning setting. Future work includes adding an estimator that divides the estimation interval according to a user-defined distance and dynamically adapting the estimator setting during runtime. It also seems relevant to compare against different learning systems with a number of probabilistically annotated datasets in order to assess the quality of the models and execution time.

Acknowledgments. Joana Côrte-Real is funded by the FCT grant SFRH/BD/52235/2013. This work is partially funded by the ERDF through the COMPETE 2020 Programme within project POCI-01-0145-FEDER-006961, by National Funds through the FCT as part of project UID/EEA/50014/2013, and by the North Portugal Regional Operational Programme, under the PORTUGAL 2020 Partnership Agreement, and through the European Regional Development Fund as part of project NanoSTIMA (NORTE-01-0145-FEDER-000016).

References

1. Bellodi, E., Riguzzi, F.: Structure learning of probabilistic logic programs by searching the clause space. Theor. Pract. Log. Program. **15**(02), 169–212 (2015)
2. Côrte-Real, J., Mantadelis, T., Dutra, I., Rocha, R., Burnside, E.: SkILL - a stochastic inductive logic learner. In: Proceedings of the 14th International Conference on Machine Learning and Applications, pp. 555–558. IEEE (2015)
3. De Raedt, L., Dries, A., Thon, I., Van den Broeck, G., Verbeke, M.: Inducing probabilistic relational rules from probabilistic examples. In: International Joint Conference on Artificial Intelligence, pp. 1835–1843. AAAI Press (2015)
4. Raedt, L., Kersting, K.: Probabilistic inductive logic programming. In: Raedt, L., Frasconi, P., Kersting, K., Muggleton, S. (eds.) Probabilistic Inductive Logic Programming. LNCS, vol. 4911, pp. 1–27. Springer, Heidelberg (2008). doi:10.1007/978-3-540-78652-8_1
5. De Raedt, L., Kimmig, A.: Probabilistic (logic) programming concepts. Mach. Learn. **100**(1), 5–47 (2015)
6. Raedt, L., Thon, I.: Probabilistic rule learning. In: Frasconi, P., Lisi, F.A. (eds.) ILP 2010. LNCS, vol. 6489, pp. 47–58. Springer, Heidelberg (2011). doi:10.1007/978-3-642-21295-6_9
7. Getoor, L.: Introduction to Statistical Relational Learning. MIT press, Cambridge (2007)
8. Getoor, L., Taskar, B., Koller, D.: Selectivity estimation using probabilistic models. In: ACM SIGMOD Record, vol. 30, pp. 461–472. ACM (2001)
9. Kersting, K., De Raedt, L., Kramer, S.: Interpreting Bayesian logic programs. In: AAAI Workshop on Learning Statistical Models from Relational Data, pp. 29–35 (2000)
10. Kimmig, A., Demoen, B., De Raedt, L., Costa, V.S., Rocha, R.: On the implementation of the probabilistic logic programming language ProbLog. Theor. Pract. Log. Program. **11**(2 & 3), 235–262 (2011)

11. Kok, S., Domingos, P.: Learning the structure of markov logic networks. In: International Conference on Machine Learning, pp. 441–448. ACM (2005)
12. Muggleton, S.: Stochastic logic programs. Adv. Inductive Log. Program. **32**, 254–264 (1996)
13. Natarajan, S., Khot, T., Kersting, K., Gutmann, B., Shavlik, J.: Gradient-based boosting for statistical relational learning: the relational dependency network case. Mach. Learn. **86**(1), 25–56 (2012)
14. Richardson, M., Domingos, P.: Markov logic networks. Mach. Learn. **62**(1–2), 107–136 (2006)
15. Costa, V.S., Page, D., Qazi, M., Cussens, J.: CLP(BN): constraint logic programming for probabilistic knowledge. In: Conference on Uncertainty in Artificial Intelligence, pp. 517–524 (2002)
16. Sato, T., Kameya, Y.: PRISM: a language for symbolic-statistical modeling. In: International Joint Conference on Artificial Intelligence, vol. 97, pp. 1330–1339. Morgan Kaufmann (1997)
17. Schulte, O., Khosravi, H., Kirkpatrick, A., Gao, T., Zhu, Y.: Modelling relational statistics with Bayes nets. Mach. Learn. **94**(1), 105–125 (2014)

Inductive Logic Programming Meets Relational Databases: Efficient Learning of Markov Logic Networks

Marcin Malec[1], Tushar Khot[2], James Nagy[3], Erik Blask[3],
and Sriraam Natarajan[1(✉)]

[1] Indiana University Bloomington, Bloomington, IN, USA
{mmalec,natarasr}@indiana.edu
[2] Allen Institute of AI, Seattle, USA
[3] Air Force Research Laboratory, Riverside, USA

Abstract. Statistical Relational Learning (SRL) approaches have been developed to learn in presence of noisy relational data by combining probability theory with first order logic. While powerful, most learning approaches for these models do not scale well to large datasets. While advances have been made on using relational databases with SRL models [14], they have not been extended to handle the complex model learning (structure learning task). We present a scalable structure learning approach that combines the benefits of relational databases with search strategies that employ rich inductive bias from Inductive Logic Programming. We empirically show the benefits of our approach on boosted structure learning for Markov Logic Networks.

1 Introduction

Recently, a great deal of progress has been made in developing (probabilistic) methods that can directly learn from relational data, in what is now known as Statistical Relational Learning (SRL) or Probabilistic Logic Models (PLMs) [6,18]. The advantage of PLMs is that they can succinctly represent probabilistic dependencies among the attributes of different related objects, leading to a compact representation of learned models while effectively modeling uncertainty.

While the combination is potent from a representation perspective, learning is expensive. In particular, we consider the formalism of Markov Logic Networks where model learning has been pursued actively in recent times [1,8,9]. The key issue is the fact that as with standard Inductive Logic Programming search different levels of abstractions (populations, sub-populations, individual objects) must be explored. In addition, the weights need to be fixed for every clause induced. Hence, many of the resulting approaches make limited assumptions to facilitate effective model learning. Some of these restrictions include the finite domain assumption (Herbrand interpretations)[1], not allowing for functor

---

[1] Some models such as Blog [12] allow for relaxing these assumptions but as far as we are aware, they do not have a full model learning algorithm.

© Springer International Publishing AG 2017
J. Cussens and A. Russo (Eds.): ILP 2016, LNAI 10326, pp. 14–26, 2017.
DOI: 10.1007/978-3-319-63342-8_2

symbols (i.e., learning only using predicates), not allowing for recursion etc. In essence, most of these methods mainly exploit "parameter tying" i.e., allowing for instances of objects to share the same parameters under the same conditions.

Consequently, PLM systems have been built using relational databases [19]. For example, more recently, a probabilistic database system called Tuffy [14], has been developed for a particular SRL model called Markov Logic network [4]. It is an efficient database implementation that employs PostgreSQL underneath. This system has shown to have efficient parameter learning (learning of weights) and has been extended to general factor graph learning [15]. However, these systems are restricted to learning only the parameters of the underlying models (weights/probabilities/potential functions) and not the full model (rules/structure of graphical models).

Our hypothesis is that these data base systems can benefit from advances inside ILP [10]. Recall that most systems employ additional directives, typically called *modes*, to restrict the search space such that the learning of these clauses is efficient. We propose to employ the success of ILP methods inside relational databases to accelerate the full model learning of SRL models. Inspired by the recent work on QuickFOIL [21], we employ the use of background knowledge inside the database system used by Tuffy. The key difference to QuickFOIL is that we are not just learning a set of rules but a set of weighted rules. To this effect, we adapt the state-of-the-art MLN learning algorithm based on functional-gradient boosting [7]. This boosting method has been shown to be effectively learning MLNs across several domains and employs the use of modes to guide the search space. We show that combining the scalability of a relational database system with the effectiveness of mode-directed ILP learning will result in huge performance gains compared to the best learning system.

We make the following key contributions: we consider the task of learning SRL models effectively and propose a database solution for this task. We demonstrate how the efficiency and effectiveness of the search space can be improved by using background knowledge inside databases. We consider a powerful learning algorithm and show how it can be further improved by the use of databases. Finally, we demonstrate empirically that the proposed ideas outperform the baseline methods on several benchmark data sets.

2   Background

We first define some notations that will be used in this work. We use capital letters such as X, Y, and Z to represent random variables (atoms in our formalism). We use small letters such as x, y, and z to represent values taken by the variables and bold-faced letters to represent sets.

Markov Logic Networks. A Markov Logic Network consists of a set of formulas in first-order logic and their real-valued weights, $\{(w_i, f_i)\}$. Each grounding of a clause corresponds to a factor with the potential function $\exp(w_i)$, leading to the joint probability distribution, $P(\mathbf{x}) = \frac{1}{Z} \exp\left(\sum_i w_i n_i(\mathbf{x})\right)$, where $n_i(\mathbf{x})$ is the number of times the ith formula is satisfied by $\mathbf{x}$ and Z is the normalization constant. The weights of the rule can be interpreted as weights in Markov networks, i.e., higher the weights, more likely is the rule to be true. Due to the exponential size of the normalization constant, most learning approaches maximize the pseudo-loglikelihood given as $PLL(\mathbf{X} = \mathbf{x}) = \sum_i \log P(X_i = x_i \mid MB(x_i))$ where $MB(x_i)$ is the Markov blanket of x_i.

Boosting MLNs. We employ relational functional gradient boosting (RFGB) approach developed for MLNs [7]. RFGB approach like Friedman's boosting [5], performs gradient ascent on the functional space. To do so, the probability distribution of each relational example, $P(x_i \mid MB(x_i))$ is represented as a sigmoid over a regression function $\psi(x_i; \mathbf{MB}(x_i))$. The gradients can be computed on the pseudo-loglikelihood function w.r.t. the function ψ as $\frac{\partial PLL(\mathbf{X}=\mathbf{x})}{\partial \psi(x_i;\mathbf{MB}(x_i))} = I(x_i = 1) - P(x_i = 1; \mathbf{MB}(x_i))$ which is the difference between the true distribution (I is the indicator function) and the current predicted distribution. Hence these gradients are positive for positive examples and negative for negative examples. RFGB starts with an initial function ψ_0 defined over all the relational examples (ground atoms) and computes the gradients for all the examples, Δ_1. A regression function, $h_1 : X \to \mathbb{R}$ is then learned to fit to these gradients and added to the initial function i.e. $\psi_1 = \psi_0 + h_1$. This process is repeated n times and the final ψ function for an example is given as the sum of values from all the gradient functions, $\psi_n(x) = \psi_0(x) + h_1(x) + \cdots + h_n(x)$.

For MLNs, the regression function is $\psi(x_i; \mathbf{MB}(x_i)) = \sum_j w_j nt_j$ $(x_i; \mathbf{MB}(x_i))$ where $nt_j(x_i; \mathbf{MB}(x_i))$ corresponds to the non-trivial groundings [20] of an example x_i given its Markov blanket , $nt_j(x_i; \mathbf{MB}(x_i)) = n_j(x_i = 1, \mathbf{MB}(x_i)) - n_j(x_i = 0, \mathbf{MB}(x_i))$. Relational regression trees or clauses can now be learned to fit to these gradients. We focus on the learning regression clauses. Thus, each gradient step (h_n) is a regression clause and the final model $\psi_n(x) = \psi_0(x) + h_1(x) + \cdots + h_n(x)$ is a sum over the values returned by the regression clauses. Note that learning these clauses would require computing the number of groundings for every candidate clause.

Modes in ILP. A mode definition for a predicate determines whether a particular literal, say p(X) will be considered for addition to a clause. The three types of modes considered here are:

- p(+) : the variable used as p's argument must already appear in the clause. E.g. p(X) and p(Y) would be considered for addition to q(Y) :- r(X, Y).
- p(-) : the variable used as p's argument need not appear in the clause. E.g. p(X), p(Y) and p(Z) would be considered for addition to q(Y) :- r(X, Y).
- p(#) : p's argument needs to be a constant. E.g. $p(c_1),...p(c_n)$ would be considered for addition to q(Y) :- r(X, Y).

3 Learning Statistical Relational Models Using Databases

We now present our proposed framework where we employ the use of in-memory databases for learning relational rules with their parameters. First, we describe the problem and then show how each component, that of specifying the background knowledge, the search over the space of hypothesis and the boosting process itself is performed in the databases. We provide a standard SRL example of smokes and cancer as a running illustration.

3.1 Problem Description

Given: Background knowledge (B), a set of propositional facts – evidence (F), a set of positive (P) and negative examples (N) for a set of target predicates.

    To Do: Use in-memory database to learn a discriminative MLN via RFGB.

    Output: The set of learned weighed logic rules (horn clauses).

    We used the database engine HyperSQL (HSQLDB) in embedded mode. We will consider the following running example throughout the paper.

    Illustrative Example: We consider the classic *smokers-friends-cancer* example [4] which has facts about who smokes, and the list of friends. The goal is to predict who will have cancer based on smoking status and social relationships.

3.2 Encoding Background Knowledge

Recall that background knowledge of ILP consists of two components:

- Predicate definitions - the names of the predicates and the specification of the domains for the predicate's arguments
- Mode definitions - the rules for the predicate arguments in a candidate literal.

The modes serve to restrict the language and acts as an inductive bias to the search process. Recall that our current system is inspired from the MLN boosting method [7], a discriminative learning approach. The goal is to learn a set of horn clauses and the modes essentially serve to describe the predicates in the hypothesis Horn clauses. An important use of modes is that they serve to restrict the use of existentially quantified variables in the learned horn clauses.

Illustrative Example: Returning to smokes-cancer example, the background file declaration in logic format could look as follows:

```
predDef: friends(person, person).
predDef: smokes(person).
predDef: cancer(person).
mode: friends(+, -).
mode: friends(-, +).
mode: smokes(+).
mode: cancer(+).
```

Fig. 1. Mode search space reduction. (Color figure online)

As with standard ILP systems, the use of modes in our learning algorithm can be clearly seen in Fig. 1. The current learning task is to predict $Cancer(X)$ (green node in the center). The modes restrict our next expansion search space to the nodes shown in green. As can be seen due to the use of + in Smokes predicate, we only consider $Smokes(X)$ for expansion and not a new existential variable say $Smokes(Y)$. Similarly, some of the friends of X must be introduced into the search space before considering their friends and their smoking habits. These constraints are key for ILP systems to work efficiently and we adapt them in the context of learning with databases.

3.3 Facts

We now show how the facts and the positive and negative examples are encoded in our work. Following prior work in SRL, we make the *closed-world assumption*, i.e., all the groundings that are not specified in the fact base (unobserved groundings) are false. All the true facts are stored in the database with each predicate corresponding to one table and each argument of the predicate corresponding to a column in the table.

In the case of target predicates we use an additional column that contains the truth value of the grounding. Since we are learning a MLN, the MLN semantics requires us compute $PSUM$ ($\Sigma_i SATcount_i(x) \times clauseWeight_i$) for each example which is stored as an additional column. This is essentially a sum over the weighted count of the number of satisfied groundings of each clause. Recall that we are performing functional gradient descent, and hence we also need to compute the gradients ($Truth\text{-}value - sigmoid(PSUM)$) for each example. Finally, given the need to compute the difference between the number of satisfied and unsatisfied groundings in the gradient, we also store the negative facts. In our experiments, $PSUM$ is initialized to -1.8 (as an initial prior as it was suggested in the work of Khot et al. [7]). In the next section, we show how the facts and background knowledge of the smokers example is fully encoded in our database.

Illustrative Example: Let us consider the task of predicting cancer. Let the true facts for this domain be as follows:

smokes(chuck) friends(bob, chuck) cancer(bob)
smokes(bob)   friends(bob, dan)   cancer(chuck)
              friends(chuck, bob) cancer(fred))
              friends(chuck, fred)
              friends(dan, bob)
              friends(fred, chuck)

These facts would be stored inside the database as shown in Fig. 2 (left). As can be seen, the groundings of the *Cancer* predicate (which is the query predicate) are stored as a table with the log priors given as PSUM. The gradients are essentially the initial values based on the priors and these are stored in the table as well. They will be modified through the learning process with the aim of driving them to 0.

atom_Cancer

Truth	PSUM	G	ARG0
1	-1.800	0.858	bob
1	-1.800	0.858	chuck
1	-1.800	0.858	fred
0	-1.800	-0.142	dan

atom_Friends

ARG0	ARG1
bob	chuck
bob	dan
chuck	bob
chuck	fred
dan	bob
fred	chuck

atom_Smokes

ARG0
chuck
bob

Fig. 2. Representation of facts and positive examples in data bases.

Given that the positive and negative examples are stored as tables, now the rest of the facts are captured using the friends and smokes tables as shown in Fig. 2 (center & right). Finally, the gradient G is computed using the query:

```
Update atom_Cancer SET G = truth - (1.0 / (1.0 + exp(-PSUM)))
```

This is the initial value of the gradient which is computed using the truth value (1 for true and 0 for false grounding) and the prior weight (PSUM). We now turn our attention to implementing the ILP search.

3.4   ILP Search Using Databases

The search begins with a horn clause with head being the target. The database representation of the initial clause would consist of a view K that corresponds to the groundings of the initial clause with column names changed to variables.

The next step is to calculate the score of the clause. This is one of the steps where querying a database can be extremely useful. First, we filter out clauses that cover too many or too few examples as they would be not discriminative. In our experiments, we filtered clauses that covered or ignored 97.5% of the examples. For the accepted clauses, a table I is created which contains positive satisfiability counts for the groundings of the head atom. The entries in the table are populated using the following query:

`Select count(*), head's vars group by head's vars`

To compute the weight we would join the *I* table with the target table to link the gradient values, and then do the computation using aggregate functions:

`weight = Select sum(G * SAT) / sum(SAT * SAT) FROM I inner join atom_target on var1 = arg0 ...`

The next step would be to compute the score using an outer join:

`score =- Select sum(Power((SAT * weight - G), 2)) FROM I right outer join atom_target on var1 = arg0...`

Illustrative Example: Returning to the task of modeling cancer, to expand the initial clause to include Smokes(X), we use the following queries:

`Entries in I table: Select count(*), var1 group by var1`

`weight = Select sum(G * SAT) / sum(SAT * SAT) FROM I inner join atom_cancer on var1 = arg0`

`score =- Select sum(Power((SAT * weight - G), 2)) FROM I right outer join atom_cancer on var1 = arg0`

The entries in the I table are then: *I* table

SAT	var1
1	bob
1	chuck

This process would be repeated for every candidate literal, and then for each of the resulting clauses limited using beam search. The best clause found using such search would then be added to the model. Once a clause is added to the model its I table's SAT counts and clausal weight are used to update the PSUM values of the head's atom table. Then the gradient values are recomputed.

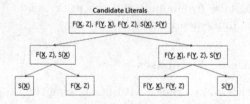

Fig. 3. Use of partitions.

Use of Modes: To generate the reduced set of candidate literals all combination of atoms are generated with restriction that domain of each predicate argument is limited to existing variable if + is specified, and existing variable and possible new variables if − specified, or constants if # is specified. These are stored in a set to eliminate duplicates. For the cancer task, the candidate literals considered in the first gradient step would only include

$\langle Smokers(X), Friends(X,Y), Friends(Y,X) \rangle$. To speed-up the search each gradient step is limited to expanding only 10 best clauses in each gradient step. Finally, the SAT counts remain the same across gradient iterations, so the I tables are not reused if the same clause is to be reevaluated.

The conversion to the database format allows for efficient query and retrieval of the data. This in turn allows for counting the satisfied groundings of any clause efficiently. As has been shown before [17], counting the satisfied grounding is the bottleneck in many PLM tasks including learning and inference. Efficient grounding could possible allow for improving the speed of these tasks.

It must be mentioned that our efficiency does have some limitations - (1) we assume a finite set of groundings (possibly a large set but a finite set). (2) Only horn clauses can be learned using our method and (3) We make the

```
Function MLN Boost(Data)
    for 1 ≤ m ≤ M do
        F_m := F_{m-1}
        for P in T do
            S := GenExamples(Data; F_{m-1}, P)
            Δm := FitRelRegressClauseDB(S, P, N, B)
            F_m := F_m + Δm
        end
    end
Function FitRelRegressionClauseDB((S, P, N, B))
    Beam := {P(X)}
    BC := P(X)
    while ¬ empty(Beam) do
        Clause := popFront(Beam)
        if length(Clause) ≥ N then
            continue
        end
        C := getCandidateLiterals(Clause)
        Q := getPartitions(C)
        QCounts = getCountsUsingJoins(Q, Clause)
        CCounts := evaluateClauses(P, C, Counts)
        for c ∈ C do
            c.score = SE(c, CCounts(c), S)
            if c.score ≥ Clause.score then
                insert(Beam, c, c.score)
            end
            if c.score ≥ BC.score then
                BC := c
            end
        end
        while length(Beam) ≥ B do
            popBack(Beam)
        end
    end
    return BC
```

Algorithm 1. MLN-Boost Algorithm

closed-world assumption to perform counting efficiently. However, we argue and show empirically that these assumptions are practically useful in many PLMs. Particularly, the state-of-the-art learning method for MLNs make these assumptions but is built on a java-based system. We replace the java system with our database system and show significant efficiency gains without losing the performance.

Partitioning Candidate Literals: We partition the candidate literals into groups in which members of the same group share a common join. The idea is to do the shared join only once to speed up the learning time. An example of partitioning is shown in Fig. 3.

Algorithm for Learning MLNs: Algorithm 1 describes our approach applied to boosting MLNs [7]. MLN Boost function presents the boosting approach as described by Khot et al. [7]. We first generate the regression examples based on the gradients described earlier and learn regression clauses to fit these gradients. We change the regression clause learner to use our database representation in FitRelRegressionClauseDB.

We use the standard beam search to search over the space of candidate clauses. The parameter N specifies the maximum length of the learned clauses (set to 3 in our experiments) and B specifies the beam size (set to 10). To compute the score of the candidate literals, we first compute the partitions of the literals being considered in getPartitions. We use database queries to get the counts of the partitions joined with the current clause in getCountsUsingJoins. Finally given these counts over the partitions, we can compute the counts of each example for every candidate literal (evaluateClauses). These counts can then be used to compute the squared error (SE) while scoring literals during search.

4   Empirical Evaluation

We now present the results of using our approach on standard benchmark SRL data sets. We aim to evaluate the following questions:

- **(Q1)** Does the proposed database based SRL system outperform the baseline in terms of learning time?
- **(Q2)** Does the proposed system sacrifice learning performance for efficiency?

Since we are in relational domains, it is well-known that most of the relations are false - i.e., negative examples far outnumber the number of positives. In such cases, it has been frequently observed that other measures such as Area under the Precision-Recall curve (AUC-PR), Area under Receiver Operating Characteristic curve (AUC-ROC) are considered more reasonable alternatives. Hence, we primarily focus on three performance measures - AUC-ROC and AUC-PR for measuring the performance efficacy and the time in seconds for measuring efficiency. Further, for Cora, IMDB and WebKB datasets we have subsampled the negative examples at each gradient step during learning to be twice in number as the number of the positive examples. Our hypothesis is that our system can match the state-of-the-art learning algorithm in learning an accurate model in significantly faster time. We consider the following approaches:

1. BoostR - WILL based MLN boost algorithm, that serves as our reliable baseline.
2. DB_Boost_NM - Database powered MLN boost without modes, that serves as our DB baseline. This system searches exhaustively for the horn clauses.
3. DB_Boost - Database powered MLN boost that caches join results.

	AUC-ROC	AUC-PR	Time(s)
Smokers			
BoostR	1.0	1.0	2.002
DB_Boost_NM	0.5	0.6	2.196
DB_Boost	1.0	1.0	0.376

Smokers is a popular synthetic testbed that is used by several SRL methods for evaluation [4,7,13]. It consists of 3 predicates: Smokes, Friends, and Cancer. We chose cancer to be our target, our train domain had 6 people, and our test domain had 8 people. Being a small domain, we do not expect significant improvement in run times. However, as can be observed, the database boosting method that uses modes is still thrice as fast as the baseline method with the same AUC.

	AUC-ROC	AUC-PR	Time(s)
Cora Entity Resolution			
BoostR	0.521	0.141	804.877
DB_Boost_NM	-	-	> 7200
DB_Boost	0.511	0.157	13.030

The Cora dataset, now a standard dataset for citation matching, was first created by Andrew McCallum, later segmented by Bilenko and Mooney [2], and fixed by Poon and Domingos [16]. In citation matching, a group is a set of citations that refer to the same paper, and a nontrivial group contains more than one citation [16]. The Cora dataset has 1,295 citations and 134 groups where almost every citation in Cora belongs to a nontrivial group; the largest group contains 54 citations. It contains the predicates: HasWordAuthor, HasWordTitle, HasWordVenue, Title, Venue, Author.

We performed 5-fold cross-validation, and we record average time over the 5 folds. Without the use of modes the database boost algorithm search was not making much progress and we have terminated it at 2 h. As with the previous experiments, it can be observed that the learned models of our approach exhibit the same prediction performance with databases as that of the original BoostR system. This answers Q2 by showing that we do not sacrifice learning performance while still being significantly faster than the original system.

	AUC-ROC	AUC-PR	Time(s)
IMDB			
BoostR	0.986	0.527	27.741
DB_Boost_NM	0.508	0.147	4525.743
DB_Boost	0.985	0.513	3.432

The IMDB dataset was first used by Mihalkova and Mooney [11] and contains five predicates: actor, director, genre, gender and workedUnder. Since gender can take only two values, we convert the gender(person, gender) predicate to a single argument predicate `female_gender(person)`. Following prior work [7], we omitted the four equality predicates. We performed five-fold cross-validation using the folds generated by Mihalkova and Mooney to build model for the target workedUnder and we record average time over the 5 folds.

In this data set, both systems achieve comparable AUC-ROC. However, the database based system seem to have a significantly higher AUC-PR. This is due to improved recall. Investigating the cause of this improvement is an important research direction. In terms of learning time, both systems are fast. However, the proposed system is still marginally faster than the original boostR system.

		AUC-ROC	AUC-PR	Time(s)
WebKB	BoostR	0.932	0.038	4.161
	DB_Boost_NM	-	-	> 7200
	DB_Boost	0.936	0.039	1.221

The WebKB dataset was first created by Craven et al. [3] and contains information about department webpages and the links between them. It also contains the categories for each webpage and the words within each page. This dataset was converted by Mihalkova and Mooney [11] to contain only the category of each webpage and links between these pages. They created the following predicates: Student(A), Faculty(A), CourseTA(C, A), CourseProf(C, A), Project(P, A) and SamePerson(A, B) from these webpages. The textual information was ignored. We removed the SamePerson(A, B) predicate as it only had groundings with both the arguments being exactly same (i.e., SamePerson(A,A)). We evaluated our method over the CourseProf predicate. We performed 4-fold cross-validation where each fold corresponds to one university, and we record average time over the 4 folds. Without the use of modes the database boost algorithm search was not making much progress and we have terminated it at 2 h. It can be observed that the AUC-ROC and AUC-PR are comparable with the BoostR system for the different database systems. However, the proposed system is significantly faster than the original while learning a comparable model.

Discussion: In summary, it can be **clearly observed** that the proposed database based systems that uses modes are significantly faster than the original BoostR system. However, this performance is achieved without significantly losing learning accuracy. Hence, **Q1** can be answered affirmatively in that the proposed methods are significantly faster than the state-of-the-art baseline. **Q2** can be answered negatively in that we do not sacrifice learning performance for improved learning time.

5   Conclusion and Future Work

We considered the problem of scaling up a successful boosting algorithm for SRL models. To this effect, we designed a in-memory database solution that exploited the search bias used in many logical models. Our initial evaluations clearly demonstrate that this learning system is capable of learning accurate models in significantly shorter amount of time. Extensive evaluations of this approach is our next immediate direction for future research. Employing approximate counts for the groundings will potentially allow for even greater savings in time. However, these approximations need to be theoretically analyzed for the learning performance, another interesting research direction. Finally, embedding the powerful learning approach such as boosting inside a large-scale system such as DeepDive will allow us to fully realize the gains attained in related fields.

Acknowledgements. MM and SN acknowledge the support of the DARPA DEFT Program under the Air Force Research Laboratory (AFRL) prime contract no. FA8750-13-2-0039. Any opinions, findings, and conclusion or recommendations expressed in this material are those of the authors and do not necessarily reflect the view of the DARPA, ARO, AFRL, or the US government.

References

1. Biba, M., Ferilli, S., Esposito, F.: Structure learning of Markov logic networks through iterated local search. In: ECAI (2008)
2. Bilenko, M., Mooney, R.: Adaptive duplicate detection using learnable string similarity measures. In: KDD (2003)
3. Craven, M., DiPasquo, D., Freitag, D., McCallum, A., Mitchell, T., Nigam, K., Slattery, S.: Learning to extract symbolic knowledge from the World Wide Web. In: AAAI, pp. 509–516 (1998)
4. Domingos, P., Lowd, D.: Markov Logic: An Interface Layer for AI. Morgan & Claypool, San Rafael (2009)
5. Friedman, J.: Greedy function approximation: a gradient boosting machine. Ann. Stat. **29** (2001)
6. Getoor, L., Taskar, B.: Introduction to Statistical Relational Learning. MIT Press, Cambridge (2007)
7. Khot, T., Natarajan, S., Kersting, K., Shavlik, J.: Learning Markov logic networks via functional gradient boosting. In: ICDM (2011)
8. Kok, S., Domingos, P.: Learning Markov logic network structure via hypergraph lifting. In: ICML (2009)
9. Kok, S., Domingos, P.: Learning Markov logic networks using structural motifs. In: ICML (2010)
10. Lavrac, N., Dzeroski, S.: Inductive Logic Programming - Techniques and Applications. Ellis Horwood Series in Artificial Intelligence. Ellis Horwood, New York (1994)
11. Mihalkova, L., Huynh, T., Mooney, R.: Mapping and revising Markov logic networks for transfer learning. In: Proceedings of the 22nd National Conference on Artificial Intelligence, vol. 1 (2007)
12. Milch, B., Marthi, B., Russell, S.: Blog: Relational modeling with unknown objects. In: Proceedings of the SRL Workshop in ICML (2004)
13. Natarajan, S., Khot, T., Kersting, K., Gutmann, B., Shavlik, J.: Gradient-based boosting for statistical relational learning: the relational dependency network case. MLJ **86**, 25–56 (2012)
14. Niu, F., Zhang, C., Re, C., Shavlik, J.: Scaling inference for Markov logic via dual decomposition. In: ICDM, pp. 1032–1037 (2012)

15. Niu, F., Zhang, C., Ré, C., Shavlik, J.: Deepdive: web-scale knowledge-base construction using statistical learning and inference. In: Second International Workshop on Searching and Integrating New Web Data Sources (2012)
16. Poon, H., Domingos, P.: Joint inference in information extraction. In: AAAI, pp. 913–918 (2007)
17. Poyrekar, S., Natarajan, S., Kersting, K.: A deeper empirical analysis of CBP algorithm: grounding is the bottleneck. In: Statistical Relational Artificial Intelligence, Papers from the 2014 AAAI Workshop, Québec City, 27 July 2014. http://www.aaai.org/ocs/index.php/WS/AAAIW14/paper/view/8776
18. Raedt, L.D., Kersting, K.: Probabilistic logic learning. SIGKDD Explor. Newsl. **5**(1), 31–48 (2003)
19. Schulte, O., Qian, Z.: SQL for SRL: structure learning inside a database system. CoRR abs/1507.00646 (2015)
20. Shavlik, J., Natarajan, S.: Speeding up inference in Markov logic networks by preprocessing to reduce the size of the resulting grounded network. In: IJCAI (2009)
21. Zeng, Q., Patel, J.M., Page, D.: Quickfoil: scalable inductive logic programming. Proc. VLDB Endow. **8**(3) (2014)

Online Structure Learning for Traffic Management

Evangelos Michelioudakis[1]([✉]), Alexander Artikis[2,1], and Georgios Paliouras[1]

[1] Institute of Informatics and Telecommunications,
NCSR "Demokritos", Agia Paraskevi, Greece
{vagmcs,a.artikis,paliourg}@iit.demokritos.gr
[2] Department of Maritime Studies, University of Piraeus, Piraeus, Greece

Abstract. Most event recognition approaches in sensor environments
are based on manually constructed patterns for detecting events, and
lack the ability to learn relational structures in the presence of uncertainty. We describe the application of OSLα, an online structure learner
for Markov Logic Networks that exploits Event Calculus axiomatizations,
to event recognition for traffic management. Our empirical evaluation is
based on large volumes of real sensor data, as well as synthetic data generated by a professional traffic micro-simulator. The experimental results
demonstrate that OSLα can effectively learn traffic congestion definitions
and, in some cases, outperform rules constructed by human experts.

Keywords: Markov Logic Networks · Event Calculus · Uncertainty

1 Introduction

Many real-world applications are characterized by both uncertainty and relational structure. Regularities in these domains are hard to identify manually,
and thus automatically learning them from data is desirable. One framework
that concerns the induction of probabilistic knowledge by combining the powers
of logic and probability is Markov Logic Networks (MLNs) [12]. Structure learning approaches that focus on MLNs have been successfully applied to a variety
of applications where uncertainty holds. However, most of these approaches are
batch and cannot handle large training sets, due to their requirement to load all
data in memory for inference in each learning iteration.

Recently, we proposed the OSLα [9] online structure learner for MLNs, which
extends OSL [7] by exploiting a given background knowledge to effectively
constrain the space of possible structures during learning. The space is con-
strained subject to the characteristics imposed by the rules governing a specific
task, herein stated as axioms. As a background knowledge we are employing
MLN–EC [14], a probabilistic variant of the Event Calculus [8,10] for event recognition.

In event recognition [1,3] the goal is to recognize *composite events* (CE)
of interest given an input stream of *simple derived events* (SDEs). CEs can be

© Springer International Publishing AG 2017
J. Cussens and A. Russo (Eds.): ILP 2016, LNAI 10326, pp. 27–39, 2017.
DOI: 10.1007/978-3-319-63342-8_3

defined as relational structures over sub-events, either CEs or SDEs, and capture the knowledge of a target application. Due to the dynamic nature of real-world applications, the CE definitions may need to be refined over time or the current knowledge may need to be enhanced with new definitions. Manual curation of event definitions is a tedious and cumbersome process, and thus machine learning techniques to automatically derive the definitions are essential.

We applied OSLα to learning definitions for traffic incidents using over 3 GiB of real data from sensors mounted on a 12 Km stretch of the Grenoble ring road, provided in the context of the SPEEDD project[1]. The goal of SPEEDD is to develop a system for proactive, event-based decision-making, where decisions are triggered by forecast events. To allow for event recognition and forecasting, OSLα is employed to construct and refine the necessary CE definitions. Due to the high volume of the dataset, the learning process must employ an online strategy. To evaluate further the predictive accuracy of OSLα, we employed a synthetic dataset generated by a professional traffic micro-simulator [13], developed by domain experts to allow for the systematic testing of the SPEEDD components.

The remainder of the paper is organized as follows. Section 2 presents OSLα, while Sect. 3 describes the application of OSLα to the traffic domain. Section 4 summarizes the presented work and outlines further research directions.

2   OSLα: An Online Structure Learner Using Background Knowledge Axiomatization

OSLα extends OSL by exploiting a given background knowledge. Figure 1 presents the components of OSLα. The background knowledge consists of the MLN–EC axioms (i.e., domain-independent rules) and an already known (possibly empty) hypothesis (i.e., set of clauses). Each axiom contains *query predicates* HoldsAt $\in \mathcal{Q}$ that consist the supervision, and *template predicates* InitiatedAt, TerminatedAt $\in \mathcal{P}$ that specify the conditions under which a CE starts and stops being recognized. The latter form the target CE definitions that we want to learn. OSLα exploits these axioms in order to create mappings of supervision predicates into template predicates and search only for explanations of these template predicates. Upon doing so, OSLα does not need to search over time sequences; instead it only needs to find appropriate bodies over the current time-point for the following definite clauses:

$$\text{InitiatedAt}(f, t) \Leftarrow \text{body}$$
$$\text{TerminatedAt}(f, t) \Leftarrow \text{body}$$

Given the MLN–EC axioms, OSLα constructs a set **T** that provides mappings of its axioms to the template predicates in $\mathcal{P}$ that appear in their bodies. For instance, axiom (1) of **T**

$$\text{HoldsAt}(f, t{+}1) \Leftarrow \text{InitiatedAt}(f, t) \wedge \text{Next}(t, t{+}1) \tag{1}$$

---

[1] https://speedd-project.eu.

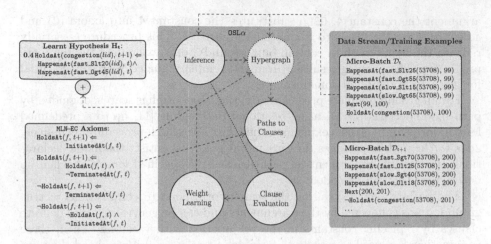

Fig. 1. The procedure of OSLα.

will be mapped to the template predicate $\texttt{InitiatedAt}(f, t)$ since the aim is to construct a rule for this predicate. The set **T** is used during the search for structures (relational pathfinding) to find an initial search set $\mathcal{I}$ of ground template predicates, and search the space of possible structures for specific bodies of the definite clauses.

At any step t of the online procedure, a training example (micro-batch) $\mathcal{D}_t$ arrives containing simple derived events (SDEs), e.g. a fast lane in a highway has average speed less than 25 km/h and sensor occupancy greater than 55%. $\mathcal{D}_t$ is used together with the already learnt hypothesis to predict the truth values $\mathbf{y}_t^P$ of the composite events (CEs) of interest. This is achieved by (maximum a posteriori) MAP inference based on LP-relaxed Integer Linear Programming [6]. Then OSLα receives the true label $\mathbf{y}_t$ and finds all ground atoms that are in $\mathbf{y}_t$ but not in $\mathbf{y}_t^P$, denoted as $\Delta\mathbf{y}_t = \mathbf{y}_t \backslash \mathbf{y}_t^P$. Hence, $\Delta\mathbf{y}_t$ contains the false positives/negatives of the inference step. Given $\mathcal{D}_t$, OSLα constructs a hypergraph that represents the space of possible structures as graph paths. Constants appear in the graph as nodes and true ground atoms as hyperedges that connect the nodes appearing as its arguments.

Then for all incorrectly predicted CEs in $\Delta\mathbf{y}_t$, OSLα uses the set **T** to find the corresponding ground template predicates for which the axioms belonging in **T** are satisfied by the current training example. Consider, for instance, that one of these is axiom (1), and that we have predicted that the ground atom $\texttt{HoldsAt(CE, 5)}$ is false (false negative). OSLα substitutes the constants of $\texttt{HoldsAt(CE, 5)}$ into axiom (1). The result of the substitution will be the following partially ground axiom:

$$\texttt{HoldsAt(CE, 5)} \Leftarrow \texttt{Next}(t, 5) \wedge \texttt{InitiatedAt(CE, } t) \tag{2}$$

Since t represents time-points and $\texttt{Next}$ describes successive time-points, there will be only one true grounding of $\texttt{Next}(t, 5)$ in the training data, having as argument the constant 4. OSLα substitutes the constant 4 into axiom (2) and adds InitiatedAt(CE, 4) to the initial search set $\mathcal{I}$. This procedure essentially reduces the hypergraph to contain only ground atoms explaining the template predicates. The pruning resulting from the template guided search is essential for learning in temporal domains.

For all ground template predicate in $\mathcal{I}$, the hypergraph is searched, guided by path mode declarations [7] using relational pathfinding [11] up to a predefined length, for definite clauses explaining the CEs. The search procedure recursively adds to the path hyperedges (i.e., ground atoms) that satisfy the mode declarations. The search ends when the path reaches the specified length or when no new hyperedges can be added.

The paths discovered during the search correspond to conjunctions of true ground atoms connected by their arguments and can be generalized into definite clauses by replacing constants in the conjunction with variables. Then, these conjunctions are used as a body to form definite clauses using as head the template predicate present in each path. The resulting set of formulas is converted into clausal normal form and evaluated.

Evaluation takes place for each clause c individually. The difference between the number of true groundings of c in the ground-truth world $(\mathbf{x}_t, \mathbf{y}_t)$ and those in predicted world $(\mathbf{x}_t, \mathbf{y}_t^P)$ is then computed (note that $\mathbf{y}_t^P$ was predicted without c). Only clauses whose difference in the number of groundings is greater than or equal to a predefined threshold μ will be added to the MLN:

$$\Delta n_c = n_c(\mathbf{x}_t, \mathbf{y}_t) - n_c(\mathbf{x}_t, \mathbf{y}_t^P) \geq \mu \tag{3}$$

The intuition behind this measure is to add to the hypothesis clauses whose coverage of the ground-truth world is significantly (according to μ) greater than that of the clauses already learned. Finally, the weights of the retained clauses are then optimized by the AdaGrad online learner [5], the weighted clauses are appended to the current hypothesis $\mathcal{H}_t$, and the procedure is repeated for the next training example $\mathcal{D}_{t+1}$.

Our implementations of OSLα, AdaGrad, and MAP inference based on LP-relaxed Integer Linear Programming, are contributed to LoMRF[2], an open-source implementation of MLNs written in Scala. LoMRF enables knowledge base compilation, parallel and optimized grounding, inference and learning.

3   Empirical Evaluation

We applied OSLα to traffic management using real data from magnetic sensors mounted on the southern part of the Grenoble ring road (Rocade Sud), that links the city of Grenoble from the south-west to the north-east. In addition to sustaining local traffic, this road has a major role, since it connects two highways: the A480, which goes from Paris and Lyon to Marseilles, and the A41, which goes

---

[2] https://github.com/anskarl/LoMRF.

from Grenoble to Switzerland. Furthermore, the mountains surrounding Grenoble prevent the development of new roads, and also have a negative impact on pollution dispersion, making the problem of traffic regulation on this road even more crucial [4]. The dataset was made available by CNRS-Grenoble, our partner in the SPEEDD project, and consists of approximately 3.3 GiB of sensor readings (one month data). Sensors are placed in 19 collection points along a 12 km stretch of the highway. Each collection point has a sensor per lane. Sensor data are collected every 15 s, recording the total number of vehicles passing through a lane, average speed and sensor occupancy. Annotations of traffic congestion are provided by human traffic controllers, but only very sparsely.

To deal with this issue, and test further OSLα, we used a synthetic dataset generated by a professional traffic micro-simulator [13], developed in the context of SPEEDD. The simulator is based on AIMSUN[3] — a widely used transport modeling software that uses a microscopic model simulating individual vehicle movement, based on statistical laws from car-following and lane-changing theories. Typically, vehicles enter a transportation network using a statistical distribution of arrivals. The microscopic model incorporates sub-models for acceleration, speed adaptation, lane-changing, etc., to describe how vehicles move, interact with each other and the infrastructure. The synthetic dataset concerns the same location — the Rocade Sud — and consists of 6 simulations of one hour each (≈18.6 MiB). The simulator has been calibrated using real traffic data. Unlike the real dataset, artificial sensors exist in 98 collection points of the highway, there is no distinction between (fast, queue, etc.) lanes, and sensor measurements additionally include vehicle density. Furthermore, the synthetic dataset is much better annotated than the real dataset.

3.1 Learning Challenges

Both datasets used for learning traffic congestions exhibit several challenges. Concerning the real dataset, the first challenge is its size, making the use of batch learners, as well as online learners such as OSL that cannot make use of background knowledge, prohibitive. For instance, in the empirical evaluation presented in [9], OSL was tested on a much smaller training set (≈2.6 MiB) and required ≈25 h to process just 40% of the data. Second, as mentioned above, traffic congestion annotation is largely incomplete, leading to the incorrect penalization of good rules. This issue is illustrated in Fig. 2.

Third, the quality of information of each sensor differs considerably. This issue is illustrated in Fig. 3, that displays the average speed of the fast lane and the queue lane at the same location, as well as the congestion annotation. Figure 3 shows that the information provided by the sensors of the queue lane is largely uninformative.

---

[3] http://www.aimsun.com/.

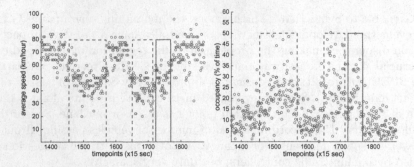

Fig. 2. Location 353708, fast lane: average speed (left) and occupancy (right). The blue points indicate the average speed (occupancy), the green windows indicate the congestion annotated by human experts, and red (dashed) windows the potentially missing annotations. (Color figure online)

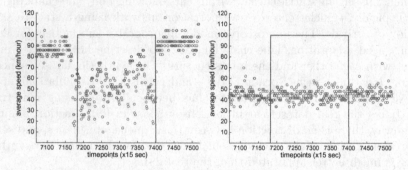

Fig. 3. Location 347549: fast lane (left) vs queue lane (right). The blue points indicate the average speed while green windows indicate the congestion annotated by human experts. (Color figure online)

Fourth, generic, location- and lane-agnostic rules are not sufficient. Consider, for example, a simple rule defining traffic congestion for any possible location regardless of the lane type:

$$\texttt{InitiatedAt}(\texttt{congestion}(lid), t) \Leftarrow$$
$$\texttt{HappensAt}(\texttt{aggr}(lid, occupancy, avgspd), t) \land$$
$$avgspd < 50 \land occupancy > 25$$

According to the above rule, a congestion in some location is said to be initiated if the average speed is below 50 km/h and the occupancy is greater than 25%. Similar rules, not shown here to save space, terminate the recognition of congestion. The optimization of the weights of these rules had large fluctuations along the learning steps, leading to zero crossings, indicating that the rules correctly capture the concept of traffic congestion in a few locations, and completely fail in others. To deal with this issue, location- and lane-specific rules must be constructed.

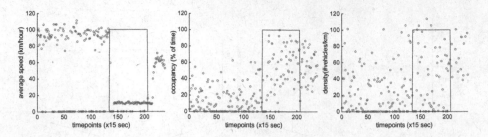

Fig. 4. Simulation 2 at location 1320: average speed (left), occupancy (middle) and density (right). The green windows indicate the congestion supervision. (Color figure online)

On the other hand, the synthetic dataset introduces different types of challenge. Density and occupancy measurements are mostly noisy, while there are a lot of zero values in all sensor measurements, leading to detection errors. These issues are illustrated in Fig. 4. Furthermore, although supervision is more complete compared to the real dataset, there are cases of missing annotation.

3.2 Experimental Setup and Results

Sensor readings constitute the simple derived events (SDEs), while traffic congestion is the target CE. The data are stored in a PostgreSQL database and the training sequence for each micro-batch, as shown in Fig. 1, is constructed dynamically by querying the database. A set of first-order logic functions is used to discretize the numerical data (speed, occupancy) and produce input events such as, for instance, HappensAt(fast_Slt55(53708), 100), representing that the speed in the fast lane of location 53708 is less than 55 km/h at time 100. The CE supervision indicates when a traffic congestion holds in a specific location. Each training sequence is composed of input SDEs (HappensAt) over the first-order logic functions and the corresponding CE annotations (HoldsAt). The total length of the training sequence in the real data case consists of 172, 799 time-points, and we consider only SDEs from fast lanes. In the synthetic data the total training sequence length consists of 238 time-points and there is no distinction between lanes.

In the experiments presented below, we compare:

- OSLα starting with an empty hypothesis.
- OSLα starting with manually constructed traffic congestion definitions developed in collaboration with domain experts.
- The AdaGrad [5] online weight learner operating on the aforementioned handcrafted definitions.

The evaluation results were obtained using MAP inference [6] and are presented in terms of F_1 score. In the real dataset, all reported statistics are micro-averaged over the instances of recognized CEs using 10-fold cross validation over

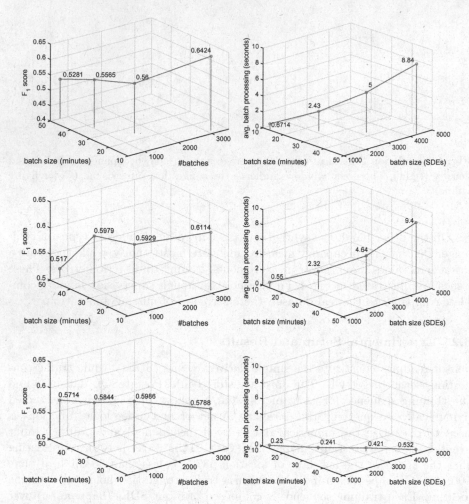

Fig. 5. Real dataset: F_1 score (left) and average batch processing time (right) for OSLα starting with an empty hypothesis (top), OSLα starting with manually constructed rules (middle) and AdaGrad operating on the manually constructed rules (bottom). In the left figures, the number of batches (see the Y axes) refers to number of learning iterations.

the entire dataset, using varying batch sizes. At each fold, an interval of 17, 280 time-points was left out and used for testing. In the synthetic data, the reported statistics are micro-averaged using 6-fold cross validation over 6 simulations by leaving one out for testing, using varying batch sizes. The experiments were performed on a computer with an Intel i7 4790@3.6 GHz processor (4 cores and 8 threads) and 16 GiB of RAM, running Ubuntu 16.04.

Real Dataset. Figure 5 presents the experimental results on the real dataset. AdaGrad and OSLα (when a starting with a non-empty hypothesis) were given location-specific rules defining traffic congestion in terms of speed and occupancy. The predictive accuracy of the learned models, both for OSLα and AdaGrad, is low. This arises mainly from the largely incomplete supervision. In OSLα, the predictive accuracy increases (almost) monotonically as the learning iterations increase. On the contrary, the accuracy of AdaGrad is more or less constant. OSLα, starting with or without the manually constructed rules, outperforms AdaGrad in terms of accuracy. (OSLα starting without (respectively with) the hand-crafted rules achieves a 0.64 (resp. 0.61) F_1 score, while the best score of AdaGrad is 0.59.) This is a notable result. The aid of human knowledge can help OSLα — see the two middle points (batch size/learning iterations) in the two top left diagrams of Fig. 5. However, OSLα achieves the best score when starting with an empty hypothesis. The absence of proper supervision penalizes the hand-crafted rules, compromising the accuracy of the learning techniques that use them. OSLα starting with an empty hypothesis is not penalized in this way, and is able to construct rules with a better fit in the data, given enough learning iterations. For some locations of the motorway, OSLα has constructed rules with different thresholds for speed and occupancy than those of the hand-crafted rules.

With respect to efficiency (see the right diagrams of Fig. 5), unsurprisingly AdaGrad is faster and scales better to the increase in the batch size. At the same time, OSLα processes data batches efficiently — for example, OSLα takes less than 10 s to process a 50-minute batch including 4, 220 SDEs.

Synthetic Dataset. To test the behavior of OSLα under better supervision, we made use of a synthetic dataset produced by a professional traffic microsimulator. The dataset concerns the same location: the southern part of the Grenoble ring road. Figure 6 presents the experimental results using only SDEs for average speed. As mentioned in Sect. 3.1, density and occupancy measurements are mostly noisy in the synthetic data. Consequently, AdaGrad and OSLα (when a starting with a non-empty hypothesis) were given rules defining traffic congestion only in terms of speed. These rules were location-agnostic since the artificial sensors do not distinguish between lanes. Not surprisingly, the predictive accuracy of the learned models in these experiments is much higher as compared to real dataset. Moreover, the accuracy of OSLα and AdaGrad is affected mostly by the batch size: accuracy increases as the batch size increases. The synthetic dataset is smaller than the real dataset and thus, as the batch size decreases, the number of learning iterations is not large enough to improve accuracy. The best performance of OSLα and AdaGrad is almost the same (approximately 0.89). In other words, OSLα starting with an empty hypothesis can match the performance of techniques taking advantage of rules crafted by human experts. This is another notable result.

The right diagrams of Fig. 6 report the average batch processing times. These diagrams verify that AdaGrad is more efficient than OSLα, and that OSLα achieves a good performance, processing batches much faster than their duration. For example, OSLα takes less than 3 s to process a 25-minute batch.

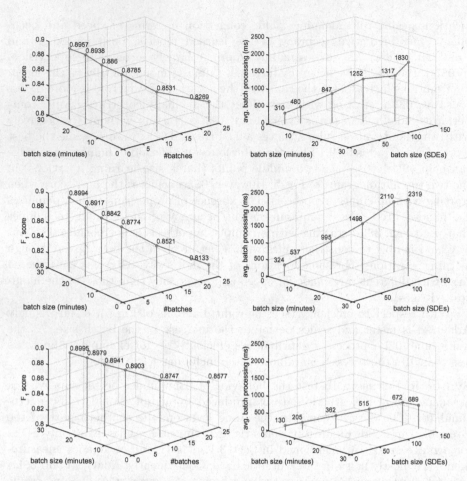

Fig. 6. Synthetic dataset with speed measurements: F_1 score (left) and average batch processing time (right) for OSLα starting with an empty hypothesis (top), OSLα starting with manually constructed rules (middle) and AdaGrad operating on the manually constructed rules (bottom).

To evaluate further the behavior of OSLα, we performed additional experiments using the synthetic dataset, this time keeping the noisy occupancy measurements. The aim of the experiments was to test OSLα in a well-annotated setting with noisy SDEs. The evaluation results are shown in Fig. 7. The hand-crafted rules defining traffic congestion combined speed with occupancy. Figure 7 shows that the noisy occupancy readings have affected the accuracy of OSLα and AdaGrad. However, OSLα starting with the manually constructed rules is affected much less, outperforming significantly both OSLα starting with an empty hypothesis and AdaGrad. OSLα has augmented the hand-crafted rules with additional clauses that focus on speed, and reduced the weight values of rules combining speed with occupancy. This way, OSLα was able to minimize the effect of noisy SDEs.

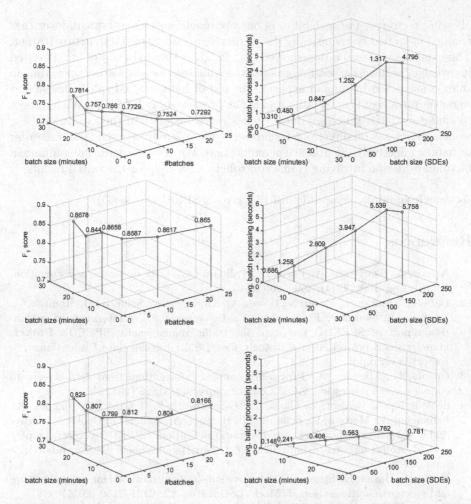

Fig. 7. Synthetic dataset with speed and occupancy measurements: F_1 score (left) and average batch processing time (right) for OSLα starting with an empty hypothesis (top), OSLα starting with manually constructed rules (middle) and AdaGrad operating on the manually constructed rules (bottom).

For completeness, the right diagrams of Fig. 7 report the average batch processing times of OSLα and AdaGrad.

4   Summary and Further Work

We presented the application of OSLα, a recently proposed structure learner for Markov Logic Networks that exploits background knowledge in the form of Event Calculus theories, to complex event recognition for traffic management. We performed an extensive empirical evaluation using over 3 GiB of real data, allowing us to test the scalability of our approach, and a synthetic dataset that enabled us to test systematically the predictive accuracy of the structure learner. The experimental evaluation showed that OSLα, without the aid of hand-crafted knowledge, performs at least as good as the AdaGrad weight learner operating on rules constructed by human experts. The aid of hand-crafted rules allows OSLα to outperform significantly AdaGrad in the presence of noisy SDEs. With respect to efficiency, OSLα processes data batches much faster than their duration.

There are several directions for further work. We aim to extend OSLα in order to handle effectively the absence of annotation. We are also performing a human factors evaluation involving traffic controllers — see [2] for the initial results.

Acknowledgments. Funded by EU FP7 project SPEEDD (619435).

References

1. Artikis, A., Skarlatidis, A., Portet, F., Paliouras, G.: Logic-based event recognition. Knowl. Eng. Rev. **27**(4), 469–506 (2012)
2. Baber, C., Starke, S., Morar, N., Howes, A., Kibangou, A., Schmitt, M., Ramesh, C., Lygeros, J., Fournier, F., Artikis, A.: Deliverable 8.5 - intermediate evaluation report of SPEEDD prototype for traffic management. SPEEDD Project. http://speedd-project.eu/sites/default/files/D8.5_-_Intermediate_Evaluation_Report-revised.pdf
3. Cugola, G., Margara, A.: Processing flows of information: from data stream to complex event processing. ACM Comput. Surv. **44**(3), 15 (2012)
4. de Wit, C.C., Bellicot, I., Garin, F., Grandinetti, P., Ladino, A., Singhal, R., Kibangou, A., Morbidi, F., Schmitt, M., Hempel, A., Baber, C., Cooke, N.: Deliverable 8.1 - user requirements and scenario definition. SPEEDD project. http://speedd-project.eu/sites/default/files/D8.1_User_Requirements_Traffic_updated_final.pdf
5. Duchi, J., Hazan, E., Singer, Y.: Adaptive subgradient methods for online learning and stochastic optimization. J. Mach. Learn. Res. **12**, 2121–2159 (2011)
6. Huynh, T.N., Mooney, R.J.: Max-margin weight learning for Markov logic networks. In: Buntine, W., Grobelnik, M., Mladenić, D., Shawe-Taylor, J. (eds.) ECML PKDD 2009. LNCS, vol. 5781, pp. 564–579. Springer, Heidelberg (2009). doi:10.1007/978-3-642-04180-8_54
7. Huynh, T.N., Mooney, R.J.: Online structure learning for Markov logic networks. Proc. ECML PKDD **2**, 81–96 (2011)
8. Kowalski, R., Sergot, M.: A logic-based calculus of events. New Gener. Comput. **4**(1), 67–95 (1986)
9. Michelioudakis, E., Skarlatidis, A., Paliouras, G., Artikis, A.: Online structure learning using background knowledge axiomatization. Proc. ECML-PKDD **1**, 237–242 (2016)
10. Mueller, E.T.: Event calculus. in handbook of knowledge representation. In: Foundations of Artificial Intelligence, vol. 3, pp. 671–708. Elsevier (2008)
11. Richards, B.L., Mooney, R.J.: Learning relations by pathfinding. In: Proceedings of AAAI, pp. 50–55. AAAI Press (1992)
12. Richardson, M., Domingos, P.M.: Markov logic networks. Mach. Learn. **62**(1–2), 107–136 (2006)

13. Singhal, R., Andreev, A., Kibangou, A.: Deliverable 8.4 - final version of micro-simulator. SPEEDD project. http://speedd-project.eu/sites/default/files/SPEEDD-D8-4_Final_Version.pdf
14. Skarlatidis, A., Paliouras, G., Artikis, A., Vouros, G.A.: Probabilistic event calculus for event recognition. ACM Trans. Comput. Log. 16(2), 11:1–11:37 (2015)

Learning Through Advice-Seeking via Transfer

Phillip Odom[1(✉)], Raksha Kumaraswamy[1], Kristian Kersting[2],
and Sriraam Natarajan[1]

[1] Indiana University, Bloomington, IN, USA
{phodom,rakkumar,natarasr}@indiana.edu
[2] Technical University of Dortmund, Dortmund, Germany
kristian.kersting@cs.tu-dortmund.de

Abstract. Experts possess vast knowledge that is typically ignored by
standard machine learning methods. This rich, relational knowledge can
be utilized to learn more robust models especially in the presence of
noisy and incomplete training data. Such experts are often domain but
not machine learning experts. Thus, deciding what knowledge to provide
is a difficult problem. Our goal is to improve the human-machine interaction by providing the expert with a machine-generated bias that can
be refined by the expert as necessary. To this effect, we propose using
transfer learning, leveraging knowledge in alternative domains, to guide
the expert to give useful advice. This knowledge is captured in the form
of first-order logic horn clauses. We demonstrate empirically the value
of the transferred knowledge, as well as the contribution of the expert
in providing initial knowledge, plus revising and directing the use of the
transferred knowledge.

1 Introduction

There has been an increased interest in building intelligent agents with a human-
in-the-loop. This interest has been partially fueled by the rapid development
of advice-taking systems [7,13,22] that do not rely merely on data but utilize
domain advice provided by the expert. While specific adaptations differ, these
systems are motivated by the fact that there has been decades of knowledge
acquired by experts in various fields and restricting them to be "mere labelers"
places undue importance on possibly noisy data while ignoring their expertise.

In this work, we consider the formalism of probabilistic logic (PL) [8] for
learning from rich, structured, and possibly noisy data. Previously, a knowledge-
based PL learning approach was proposed [19] that adapted a powerful boosting
algorithm [16] to accept human advice about specific regions of the feature/state
space to learn in structured domains. It uses a pre-defined set of human expert
rules as advice in every iteration of the boosting algorithm to ensure the learned
model is robust even in the presence of significantly noisy data.

While successful, this method assumes that the expert provides all relevant
advice in advance. This increases the burden on the expert significantly. While
in classical systems the burden on the expert was to generate examples/labels, in

© Springer International Publishing AG 2017
J. Cussens and A. Russo (Eds.): ILP 2016, LNAI 10326, pp. 40–51, 2017.
DOI: 10.1007/978-3-319-63342-8_4

knowledge-based systems the burden shifts to providing relevant advice. We aim to lessen this burden by providing the expert with an initial set of *bias* (advice rules) that the expert can modify/adapt based on their knowledge.

Inspired by a recently successful transfer learning technique that identified similarities across seemingly unrelated domains [12], we propose to employ transfer learning for providing the initial bias to the human expert. The key idea in our approach, which we call *learning through advice-seeking via transfer* (LAST), is to transfer knowledge from a source domain to generate a set of potential *advice rules* in the target domain. Then, these advice rules are provided to a domain expert who could potentially refine the current set of rules. In turn, these rules can then serve as advice for the subsequent learning algorithm.

Consider providing advice to a system that predicts the advisor of a student. The current knowledge-based PL system [19] requires the domain expert to provide rules such as "students co-author papers with their advisors", "students TA for their advisor's courses", etc. However, assume that we have knowledge in a different domain like movies where we have rules such as "actors work in movies with directors", "actors and directors typically work in similar genres", etc., to predict if an actor works with a director. Now, using the transfer learning approach, we can potentially map actors to students, directors to advisors, movies to papers, genres to departments and create a set of potentially interesting rules that can be refined by the expert in the target domain. These refined rules can then be combined with (noisy) data to learn a robust model. This can significantly reduce the burden on the expert.

This paper makes the following key contributions – (1) It proposes the first transfer-based approach for advice-giving to learning algorithms. (2) It combines a successful advice-taking PL approach and a transfer learning approach in a seamless manner. (3) It reduces the burden on the domain expert by automatically identifying relevant rules and restricts the expert input to simple refinement operations. (4) Finally, it demonstrates excellent empirical performance in several benchmark data sets and on a large real-world Never Ending Language Learning (NELL) task [3].

2 Background

Learning through advice-seeking via transfer is related to both transfer learning and knowledge-based probabilistic logic learning.

2.1 Transfer Learning

Recently, there has been an increasing interest in the development of techniques that leverage information from a possibly related task to accelerate learning in the current task. Collectively called *transfer learning* [20], they learn a model for a source task and transfer/adapt this learned model to a potentially related and similar target task. Transfer learning has been explored previously in the context of cognitive science [9,11]. Transfer learning methods that transfer across seemingly unrelated domains can be divided into two groups - the first group consists of methods that assume that the two domains share an underlying relational structure, even though they may appear dissimilar. Consequently, these methods employed higher order logic to model this structural similarity [10]. Alternatively, the second set of methods search for explicit mappings between the two domains and transfer rules from the source accordingly [14].

We consider a relational type-matching [12] transfer method called "language-bias transfer learning" (LTL) that uses type matching typically done in Inductive Logic Programming [5]. LTL utilizes types of arguments to map source predicates to target predicates, identifying similar objects in the two domains, which are then used to construct clauses in the target domain. This approach was shown to obtain state-of-the-art results in PL domains.

Inspired by this, we propose the use of this transfer method for generating good domain knowledge in the target domain that can potentially be refined by an expert. Such a generation has two major advantages. First, it reduces the burden on the expert to generate several advice clauses in the target domain that can be used for learning. Second, it improves the results of the LTL method because it allows for the model transferred by the algorithm to be used to correct (possibly noisy) data in the target domain. We now explain the background of the learning method that can effectively exploit domain advice when learning with noisy data.

2.2 Knowledge-Based Probabilistic Logic Learning

Previous work [17] extended standard functional gradient boosting [6] to learning relational models. The key intuition to functional gradient boosting is to transform the problem of discriminatively learning a large, complex model into a series of smaller, simpler problems. This is accomplished by learning a series of models—each one capturing some of the error w.r.t. the current model. In relational function gradient boosting, each step of the learning problem is to learn a single relational regression tree (RRT) [2]—binary decision trees with first-order logic atoms in the nodes and regression values in the leaves.

While shown to be effective across a wide variety of different problems, Relational Functional-Gradient Boosting (RFGB) requires high-quality data in order to learn a good model. Recently, there has been work on using knowledge-based learning to learn in the presence of noisy data in relational domains [18,19]. They introduced a knowledge-based framework that used label preferences to target and correct noisy data. It assumes that experts will have the appropriate knowledge of the domain to identify areas where noise is likely in the training data. The learning bias is specified in the form of first-order logic clauses $(\wedge f_i(x)- > advice(x))$. The body $(\wedge f_i)$ specifies a set of logical conditions that define the set of examples to which the advice will apply while the head $(advice(x))$ of the clause defines the label preferences. This can be written as a tuple $<\wedge f_i, \mathbf{Pref}_{label}, \mathbf{Avoid}_{label}>$. The aim is to learn a model that increases the probability of preferred predictions while decreasing the probability of avoided predictions.

Label preferences are a natural way for the expert to communicate. For instance, when considering heart attacks, an expert might say "people who have a family history of heart attacks are more likely to have a heart attack than a stroke". Here the body of the clause specifies that a patient's family member has had a heart attack and the head of the clause has heart attack as a preferred label to stroke.

While this framework performs well in the presence of systematic noise, it places an unnecessary burden on the expert. This is because, a key assumption is that the human expert will be able to identify the most informative advice (i.e., that the expert will know where the systematic noise is present in the data). This assumption can fail in several scenarios—for example, with domain experts who may not inspect historical data on a regular basis.

The solution that we present next is to employ transfer learning for generating the body of the advice, i.e., defining for which examples the algorithm is interested in having label preferences. The expert could then simply refine this advice based on his/her expert knowledge.

3   Advice-Seeking for Transfer

Our goal in this work is to facilitate a natural human interaction with the learning algorithm by allowing for the human to be involved in several stages—as an expert providing (minimal) knowledge in the source domain and as an expert who can potentially look at several clauses in the target domain and evaluate or refine the clauses in the target domain.

Consequently, our *Learning through Advice-Seeking for Transfer* (LAST) algorithm generates relevant advice clauses for the target domain from expert-provided clauses in a source domain. This advice serves as a recommendation for useful advice to the expert. While the expert is still responsible for providing the label preferences for the advice, his/her task is simplified through transfer learning. Thus, the significant effort required by the expert is reduced. In turn, this makes the system more cognitively aware, i.e., it thinks like a human in developing prior knowledge. It is key that the transfer algorithm is able to generate appropriate (in this case relevant, or near-relevant) knowledge in the target domain. As a motivating example, consider providing advice in the case of a movie domain.

Illustrative Example: In order to learn in a movie domain, let us assume the presence of knowledge in a university domain. This domain comprises of faculty, students, the publications that they author, and the courses in which they are involved. Networks like these are common across universities. Assume that the knowledge provided in this domain aims to predict the advisor of a student. One such piece of knowledge could be that the students are more likely to co-author with their advisor (as against with a random professor in their department).

$$paper(p1, per1), paper(p1, per2), student(per2) \Rightarrow advisor(per1, per2)$$

The goal is to use this knowledge to transfer to a movie domain (say imdb) with movies, actors and directors where the target is to predict which actors have worked under which directors. When transferring the clause from the academic domain to the movie domain, the language bias approach [12] will produce many different predictive rules. Let us consider one such clause

$$mov(m1, per1), mov(m1, per2), act(per1) \Rightarrow workedunder(per1, per2)$$

Notice how this clause captures a relationship, as actors do work under another person working on the same movie. However, this clause also covers actors working under actors in the same movie. If this rule is provided to the expert who understands the domain, he/she might suggest different refinements to this rule: (1) He/she could suggest that actors work under directors by adding a predicate to the clause. (2) Alternatively, he/she might suggest that actors do not work under each other. These lead to the following prior knowledge that can be used by the algorithm of Odom et al. [19].

$$(1) < [mov(m1, per1), mov(m1, per2), act(per1), dir(per2)],$$
$$workedunder(per1, per2), \neg workedunder(per1, per2) >$$
$$(2) < [mov(m1, per1), mov(m1, per2), act(per1), act(per2)],$$
$$\neg workedunder(per1, per2), workedunder(per1, per2) >$$

3.1 The Problem Formulation

We now formally define our problem:

Given :   Source knowledge, noisy training data, and access to an expert

Todo :   (1) Transfer knowledge from the source to target,

               (2) Solicit advice about this knowledge from expert

The goal of the advice-seeking problem is to select the transferred knowledge about which to query the expert (denoted $\mathbf{q} \subseteq \mathbf{K}$). Useful knowledge in our context is measured (δ) as a function of the performance (accuracy) on the training data ($\mathbf{D}$). While the goal is to maximize the performance of the queries on the training data, there is a cost (C) for making a query to the expert (as the expert has a limited budget to provide advice). Hence, the goal is $arg \max_{\mathbf{q} \subseteq \mathbf{K}} \delta_{\mathbf{q}}(\mathbf{D}) - C(\mathbf{q})$. We make the simplifying assumption that the algorithm can select n queries to ask the expert. We also assume that there is a constant cost for every query. It is an interesting direction to personalize these costs based on the difficulty of each query.

As mentioned earlier, we employ a relational transfer learning approach to generate knowledge ($\mathbf{K}$) in the target domain [12]. The set of queries can then be selected from the potential list of clauses by considering the performance of each clause on the training set. While the training set does suffer from noise, the hypothesis is that reasonable knowledge will still be generated and the expert can refine the knowledge into the appropriate advice.

4   The LAST Algorithm

We will now describe the components of our LAST algorithm (shown as Algorithm 1). The algorithm takes as input the noisy training data, the descriptions of the source and target domains, a set of knowledge about the source domain and a domain expert to query. The overall goal of LAST is to generate a set of knowledge about the target domain (which we refer to as advice) and use it to learn a more robust model in the target domain. The algorithm proceeds as follows:

4.1   Step 1: LTL

First, we transfer the set of knowledge in the source domain to the target using the LTL [12] algorithm. This transfer learning technique leverages the structure of the knowledge in the source domain to generate knowledge with similar structure in the target domain (making use of the domain descriptions). While effective, LTL generates all possible similar rules resulting in an impractical number of potential expert queries. The next step mitigates this issue.

4.2   Step 2: SelectBestN

We select from among the large potential set of queries by comparing their accuracy on the training set. Selecting the top N clauses allows the algorithm to control the number of queries that can be directed to the expert. While more advice benefits the learning algorithm, this parameter can be tuned based on the availability of the expert.

4.3   Step 3: Improve

While LTL can generate appropriate knowledge in the target domain, it does not account for noisy training data. Therefore, the expert can modify the knowledge as needed to correct for any differences between the source and target domains. As this can be a time consuming process, the expert is not required to correct all (or any) of the knowledge. As we show empirically, sometimes minimal refinement is all that is needed to improve the learning algorithm. In several cases, the use of a powerful learning algorithm that can exploit the provided advice as bias can be expected to perform reasonably well in the target domain with noisy data.

4.4 Step 4: Query

Now that appropriate knowledge has been provided in the target, the expert is queried about the label preferences. As mentioned in the background, advice consists of three components: the features describing to which examples the advice will apply, the preferred (more likely) labels for those examples, and the avoided (less likely) labels for those examples. The previous step defined the features while this step solicits the label preferences. Now the advice is fully defined.

4.5 Step 5: Learn

The final step learns a robust model from the noisy training data and the expert advice. We use an advice-based learning algorithm called KBPLL [19] which is able to effectively utilize the expert advice to learn in the presence of noisy training data. It learns by trading-off between the training data and any advice. A key advantage of this approach is that it can not only learn with any amount of advice, but it is also capable of handling conflicting advice.

Algorithm 1. The LAST algorithm: Learning through Advice-Seeking via Transfer

Data, domain description in the source domain ($\mathbf{D}_S$), domain description in the target domain ($\mathbf{D}_T$), expert (E), knowledge in the source domain ($\mathbf{K}_S$)
function LAST($\mathbf{Data}, \mathbf{D}_S, \mathbf{D}_T, \mathbf{K}_S, E$)
   $\mathbf{K}_T = \text{LTL}(\mathbf{Data}, \mathbf{D}_S, \mathbf{K}_S, \mathbf{D}_T)$
   $\mathbf{q} \in \mathbf{K}_T = \text{SELECTBESTN}(N, \mathbf{K}_T, \mathbf{Data})$
   $\mathbf{q}_r = \text{IMPROVE}(E, \mathbf{q})$
   $\mathbf{A} = \text{QUERY}(E, \mathbf{q}_r)$
   return LEARN($\mathbf{A}, \mathbf{Data}$)
end function

Algorithm 1 presents these different steps. Please recall that the expert is involved in both designing the source clauses and potentially refining the target clauses.

5 Experimental Evaluation

We aim to address the following questions:

Q1: How effectively does LAST utilize the transferred advice?
Q2: How important is the contribution of the expert?
Q3: Does LAST properly assist the expert in generating advice?
Q4: What is the quality of the advice generated?

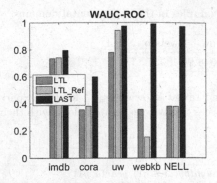
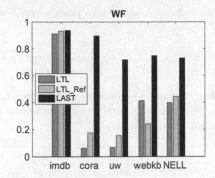

Fig. 1. Experimental results in each of our domains. We compare both weighted-AUC ROC (**LEFT**) and weighted-F score (**RIGHT**) in each domain. Higher bars represent improved performance.

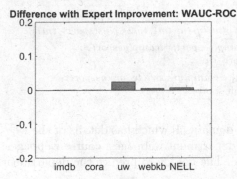
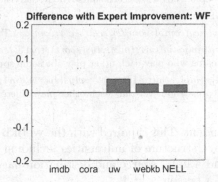

Fig. 2. Experimental results comparing the difference in performance when the expert improves the transferred advice. We compare both weighted-AUC ROC (**LEFT**) and weighted-F score (**RIGHT**). The more positive the result, the more impact the expert has in the IMPROVE step of LAST.

5.1   Domains Used

We use four standard PL domains – imdb, cora, uw and webkb – and the large-scale NELL domain. In each domain, we incorporate random noise (20%) to demonstrate that our algorithm is capable of building robust models. The domains are introduced pairwise - the source/target domain for transfer.

The **imdb** domain [15] consists of information about actors, directors and movies. The overall goal of this domain is to predict which actors work under which directors. This domain is paired with the **cora** domain [1] which consists of information about the details of author, their publications and venues. The purpose of the domain is to predict which conferences take place in the same venue.

The **uw** domain [21] consists of information about universities with details like professors, courses, students, which professor teaches which course, etc. The goal is to use this information and predict which professor advises a particular

Table 1. The negative to positive ratio of examples in the experimental domains.

Domains	imdb	cora	uw	webkb	NELL
#Neg/#Pos	14.9	1.7	548.9	400.5	7.2

Table 2. Samples of source knowledge (S) and target knowledge (T) for two experimental domains. Each domain then has two pieces of advice generated along with their English interpretation.

uw	
S: $LinkTo(c, a, b), Student(a), Dept(b)$	T: $taughtby(c1, b, q1), year(a, y1)$

$<[taughtby(c1, b, q1), year(a, y1)], advisedby, \neg advisedby>$,
Professors who teach courses advise students who are in some year of the program

$<[\neg taughtby(c1, b, q1) \lor \neg year(a, y1)], \neg advisedby, advisedby>$
People who do not teach courses do not advise students in any year of the program

NELL	
S: $acq(comp1, comp2), sect(sect1, comp2)$	T: $tmplystm(tm1, tm2), plys(sport1, tm1)$

$<[tmplystm(tm1, tm2), plys(sport1, tm1)], teamplyssport, \neg teamplyssport>$,
Teams who play each other play the same sport

$<[\neg tmplystm(tm1, tm2) \lor \neg plys(sport1, tm1)], \neg teamplyssport, teamplyssport>$
Teams who never play each other play different sports

student. This is paired with the **webkb** domain [4] which has details of the web-page structure of universities including department webpages, course webpages, and a link's source and destination page. The goal is to predict the department of a person.

NELL data is probabilistic data garnered through web crawling by an online machine learning system designed and deployed at Carnegie Mellon University [3]. We experiment with the sports domain here, where we predict what sport a particular team plays. The domain has details regarding members of a team, the sports an individual plays, and the league in which a team plays. The source used for transfer is the finance domain, where the goal is to predict to which financial sector a company belongs[1].

5.2 Evaluation Metrics Used

Relational domains naturally suffer from extreme class-imbalance as most relationships in the world are false - most students are not advised by most professors and most actors do not work with most directors. Thus, algorithms that predict all relations as false can achieve high performance on many traditional evaluation metrics. Following previous work [23], we measure performance based on weighted-AUC ROC and weighted-F scores that weigh the high recall regions of

---

[1] We do not consider transferring from finance to sports due to the lack of data from the finance domain.

ROC curve more than the low recall regions. Weighted-F score is defined as

$$F_\beta = (1 + \beta^2) \frac{Precision \times Recall}{\beta^2 \times Precision + Recall}$$

Note that the parameter β controls the trade-off between the *Precision* (percentage of correct positive predictions) and *Recall* (percentage of positive examples correctly identified). Following [23], we use $\beta = 5$. Due to the large number of negative examples in many of the domains (as shown in Table 1), the various baselines sample from the set of negative examples. Based on previous experience, LAST samples 2 to 1 (negatives to positives) while LTL and LTL_Ref sample 5 to 1.

5.3   Methods Compared

We compare LAST against several baselines. To show how LAST deals with noisy data, we compare against a language-bias transfer learning method (LTL). To show how effective the human expert is in refining and improving the advice, we compare against a variant of the previous transfer learning technique that automatically refines the transferred rules (LTL_Ref). In a separate experiment, we empirically validate the expert contribution by measuring the performance gained by the IMPROVE step of LAST.

5.4   Results

We show our results for weighted-AUC ROC (**LEFT**) and weighted-F score (**RIGHT**) in Fig. 1. In both performance measures, across all domains, our LAST algorithm performs as well as or significantly better than the baselines. More precisely, the results are significantly better than LTL and LTL_Ref except for the **imdb** domain (and **uw** for weighted-AUC ROC). As all of the methods are comparable on **imdb**, this is likely caused by this domain being easily solved (the model can be captured by a single clause). These results clearly answer (**Q1**) affirmatively, i.e., LAST is able to effectively learn with transferred advice.

As expected, the transfer learning techniques suffer from the noisy training data in several domains. In the **webkb** domain, refining the transferred knowledge (LTL_Ref) actually reduces the performance and in several other domains (**imdb**, **NELL**) there is little difference between simple transfer learning and automatically refining the rules.

The LAST algorithm has two advantages over the transfer learning baselines that allows it to effectively deal with the noisy training data. Both of these advantages directly relate to having a human-in-the-loop. Firstly, the automatic refinement (function IMPROVE from Algorithm 1) allows for improvement of rules that resemble the correct knowledge, but that have been altered by the noise. Secondly, the preference labeling (function QUERY from Algorithm 1) allows the expert to control the direction of the knowledge. If noise has reversed the meaning of the knowledge, the expert can effectively correct it. Figure 2 shows the difference in performance with and without the IMPROVE step of LAST. Blue bars represent the increase in performance with expert improvement. An increase in performance is seen in 3 of the 5 domains. These results show the impact of the expert, i.e., the expert is important (**Q2**).

In our experiments, we restrict the expert to manipulating a single piece of advice for each learned model. This effectively shows if we can get reasonable performance with minimal expert time while creating the system that can generate bias that mimics the expert. Therefore, the effort required by the expert to generate the advice is inherently low. It clearly follows that generating the advice from scratch would require more effort from the expert. Consequently, (**Q3**) can be answered affirmatively. In the future, we plan to quantify the value of transferred knowledge to the expert.

Quality of the Advice: We present sample advice for two of the domains in Table 2. Each advice includes its provenance- the rule that generated it (S), the transferred rule (T), and the two pieces of advice generated. The English interpretation of each advice rule is also provided. Note that each rule becomes two pieces of advice (and the advice is given over all possible examples in the domain). Investigating the advice in each domain, all the rules generated appear to make sense - i.e., these are rules that would possibly be naturally specified by the expert if he/she had to do without a transfer system. For example, take the advice generated for NELL which states "if a team A plays team B and team B plays a sport, then team A must also play that sport". It is clear that we are able to generate reasonable rules in all domains (**Q4**).

6 Conclusion

We presented a novel transfer-based human-in-the-loop framework for advice-giving in probabilistic logic models. Our goal was to develop a system that could provide human-like advice to a learning system based on transferring knowledge from a seemingly unrelated domain. The expert can then refine the knowledge generated by our LAST algorithm as needed. We demonstrated empirically the effectiveness of LAST in the presence of noisy training data. It showed the value of using transferred knowledge as "advice" instead of merely treating them as predictive rules in the target domain.

Our next step is to develop better heuristics to select the top rules that can be refined by the expert. Also, we aim to employ human guidance not only in the process of refinement of rules but also in the transfer process itself. More precisely, we aim to use humans to provide a bias during the search of the mapping between the two domains. Also, recall that this paper aims to discriminatively learn rules for predicting individual relations. Learning a generative model that can transfer across domains and provide a useful inductive bias in the target domain remains an interesting direction. Finally, extending this work to sequential decision making tasks can lead to the development of human-like thinking machines.

References

1. Bilenko, M., Mooney, R.: Adaptive duplicate detection using learnable string similarity measures. In: ACM SIGKDD (2003)
2. Blockeel, H.: Top-down induction of first order logical decision trees. AI Commun. **12**(1–2), 119–120 (1999)
3. Carlson, A., Betteridge, J., Kisiel, B., Settles, B., Hruschka, Jr., E., Mitchell, T.: Toward an architecture for never-ending language learning. In: AAAI (2010)
4. Craven, M., DiPasquo, D., Freitag, D., McCallum, A., Mitchell, T., Nigam, K., Slattery, S.: Learning to extract symbolic knowledge from the world wide web. In: AAAI (1998)
5. Raedt, L., Frasconi, P., Kersting, K., Muggleton, S. (eds.): Probabilistic Inductive Logic Programming. LNCS (LNAI), vol. 4911. Springer, Heidelberg (2008)
6. Friedman, J.: Greedy function approximation: a gradient boosting machine. Ann. Stat. **29**(5), 1189–1232 (2001)
7. Fung, G.M., Mangasarian, O.L., Shavlik, J.W.: Knowledge-based support vector machine classifiers. In: NIPS (2002)
8. Getoor, L., Taskar, B.: Introduction to Statistical Relational Learning. MIT Press, Cambridge (2007)
9. Gros, H., Thibaut, J.P., Sander, E.: Robustness of semantic encoding effects in a transfer task for multiple-strategy arithmetic problems. In: CogSci (2015)
10. Haaren, J., Kolobov, A., Davis, J.: Todtler: Two-order-deep transfer learning. In: AAAI (2015)
11. Jones, W., Moss, J.: Interruption-recovery training transfers to novel tasks. In: CogSci (2015)
12. Kumaraswamy, R., Odom, P., Kersting, K., Leake, D., Natarajan, S.: Transfer learning via relational type matching. In: ICDM (2015)
13. Kunapuli, G., Odom, P., Shavlik, J.W., Natarajan, S.: Guiding autonomous agents to better behaviors through human advice. In: ICDM (2013)
14. Mihalkova, L., Huynh, T., Mooney, R.: Mapping and revising markov logic networks for transfer learning. In: AAAI (2007)
15. Mihalkova, L., Mooney, R.: Bottom-up learning of Markov logic network structure. In: ICML (2007)
16. Natarajan, S., Kersting, K., Khot, T., Shavlik, J.: Boosted Statistical Relational Learners. SCS. Springer, Cham (2014)
17. Natarajan, S., Khot, T., Kersting, K., Gutmann, B., Shavlik, J.: Gradient-based boosting for statistical relational learning: the relational dependency network case. Mach. Learn. **86**, 25–56 (2012)
18. Odom, P., Bangera, V., Khot, T., Page, D., Natarajan, S.: Extracting adverse drug events from text using human advice. In: Holmes, J.H., Bellazzi, R., Sacchi, L., Peek, N. (eds.) AIME 2015. LNCS (LNAI), vol. 9105, pp. 195–204. Springer, Cham (2015). doi:10.1007/978-3-319-19551-3_26
19. Odom, P., Khot, T., Porter, R., Natarajan, S.: Knowledge-based probabilistic logic learning. In: AAAI (2015)
20. Pan, S., Yang, Q.: A survey on transfer learning. IEEE Trans. Knowl. Data Eng. **22**(10), 1345–1359 (2010)
21. Richardson, M., Domingos, P.: Markov logic networks. Mach. Learn. **62**(1–2), 107–136 (2006)
22. Towell, G.G., Shavlik, J.W.: Knowledge-based artificial neural networks. Artif. Intell. **70**(1–2), 119–165 (1994)
23. Yang, S., Khot, T., Kersting, K., Kunapuli, G., Hauser, K., Natarajan, S.: Learning from imbalanced data in relational domains: a soft margin approach. In: ICDM (2014)

How Does Predicate Invention Affect Human Comprehensibility?

Ute Schmid[1], Christina Zeller[1], Tarek Besold[2], Alireza Tamaddoni-Nezhad[3],
and Stephen Muggleton[3(✉)]

[1] University of Bamberg, Bamberg, Germany
[2] University of Bremen, Bremen, Germany
[3] Imperial College London, London, UK
`s.muggleton@imperial.ac.uk`

Abstract. During the 1980s Michie defined Machine Learning in terms
of two orthogonal axes of performance: *predictive accuracy* and *comprehensibility of generated hypotheses*. Since predictive accuracy was readily
measurable and comprehensibility not so, later definitions in the 1990s,
such as that of Mitchell, tended to use a one-dimensional approach to
Machine Learning based solely on predictive accuracy, ultimately favour-
ing statistical over symbolic Machine Learning approaches. In this paper
we provide a definition of comprehensibility of hypotheses which can
be estimated using human participant trials. We present the results of
experiments testing human comprehensibility of logic programs learned
with and without predicate invention. Results indicate that comprehensibility is affected not only by the complexity of the presented program
but also by the existence of anonymous predicate symbols.

1 Introduction

Within Artificial Intelligence (AI) *comprehensibility* of symbolic knowledge is
viewed as one of the defining factors which distinguishes logic-based representations from statistical or neural ones. However, to the authors' knowledge, no
operational criterion of comprehensibility exists in the literature. This paper
addresses the issue by introducing such a definition which is inspired by "Comprehension Tests", administered to children at primary school. Such a test comprises the presentation of a piece of text, followed by questions which probe the
child's understanding. Answers to questions in some cases may not be directly
stated, but instead inferred from the text. Once the test is scored, the degree of
the pupil's answers can be assessed numerically.

In the same fashion, our operational definition of comprehensibility is based
on presentation of a logic program to an experimental participant, who is given
time to study it, after which the score is used to assess their degree of comprehension. The detailed results of such a test can be used to identify factors in the
program which affect its comprehensibility both for individuals and for groups
of participants. The existence of an experimental methodology for testing comprehensibility has the potential to provide empirical input for improvement of

© Springer International Publishing AG 2017
J. Cussens and A. Russo (Eds.): ILP 2016, LNAI 10326, pp. 52–67, 2017.
DOI: 10.1007/978-3-319-63342-8_5

Machine Learning systems for which the generated hypotheses are intended to provide insights.

```
p(X,Y)  :- p1(X,U), p1(U,Z), p1(Z,Y).
p1(X,Y) :- father(X,Y).
p1(X,Y) :- mother(X,Y).
```
(handwritten annotations: "parent" bracketing p1 clauses, "grandgrand parent" bracketing the whole)

Fig. 1. Example of student giving meaningful names to predicate symbols.

Figure 1 provides an example of such a test in which students were asked about a given program in which some predicate names were meaningful (i.e., had publicly recognisable names like *father* and *mother*) and others were anonymous (i.e., had privately defined names like *p* and *p1*). In this case, high-scoring students often unexpectedly annotated the answer scripts to indicate the name they believed to be correct.

Given renewed interest within Inductive Logic Programming (ILP) in the use of predicate invention [1,2,7,11,12] this paper explores the effects on comprehensibility of using anonymous definitions within logic programs. Within experiments we assess students' understanding of such programs within the kinship domain. Empirical results indicate that comprehensibility is positively correlated to the degree with which new predicates produce compact descriptions. Additionally comprehensibility correlates with the degree to which participants can successfully match the presented predicate with one they are already familiar with. However, somewhat surprisingly, it is negatively correlated with the amount of time taken to inspect the definitions.

The paper is arranged as follows. In Sect. 2 we discuss existing work relevant to the paper. The framework, including relevant definitions and their relationship to experimental hypotheses is described in Sect. 3. Section 4 describes the experiments, including details of the questionnaires, experimental procedure, results and discussion. Finally in Sect. 5 we conclude the paper and discuss further work.

2 Related Work

2.1 Comprehensibility

In the late 1980s Michie [8] suggested the idea of using both comprehensibility of hypotheses and predictive accuracy as performance indicators for Machine Learning. He proposed three criteria. The *weak criterion* defines Machine Learning as occurring whenever a system generates an updated basis building on sample data for improving its performance on subsequent data. The focus is put exclusively on the prediction and problem-solving aspects. The *strong criterion* expands the weak version in a second direction, requiring the system to be able additionally to "communicate its internal updates in explicit symbolic form". Lastly his *ultra-strong criterion* additionally requires the communication of updates to be "operationally effective", in which case the user is required to understand the update and any consequences to be drawn from it. While ILP systems clearly meet the weak and strong criterion (since learning outcomes are represented as symbolic logic programs), only very limited attention has been given to checking whether the ultra-strong criterion holds, which requires testing whether the user comprehends generated hypotheses.

One Machine Learning approach which engages with issues related to comprehensibility is Argument-Based Machine Learning (ABML) [9]. ABML applies methods from argumentation in combination with a rule-learning approach. Explanations provided by domain experts concerning positive or negative arguments serve to enrich selected learning examples by being included in the learning data. Although ABML enhances the degree of explanation within a Machine Learning context, like ILP, ABML fails to pass Michie's ultra-strong test since no demonstration of user comprehensibility of learned hypotheses is guaranteed.

Issues related to comprehensibility have been gaining more wide-spread attention recently in the study of classification models [4,6]. However, while these studies emphasise the need for comprehensibility, they do not offer a definitive test of the kind provided by our definition in Sect. 3. For classification models augmented comprehensibility of the classification model promises to impact positively on the trust users have in the model's prediction. For example, in medical decision-making and in the case of unexpected system outputs, comprehensibility generally increases the acceptance of the models by users. Finally, comprehensible models can unveil new insights about the internal structure of the data or its domain of origin.

In the context of AI testing and evaluation the importance of human comprehensibility of intelligent systems has very recently been emphasised in [3]. Forbus makes a case for AI as a research endeavour being equivalent to learning how to create smart *software social organisms* which should exhibit increasing abilities to participate in human culture and daily life. The comprehensibility of the systems behaviour and outputs is paramount in this context, since only efficient communication enables participation in human society. In [3] this is then tied into the general context of assessing the capacities of AI systems by measuring the progress which has been made, for instance, in domain generality, acquired knowledge levels, or the flexibility across different interaction modalities. When looking back at the original Turing Test [19] and present-day discussions surrounding new and updated versions or substitutes for it, comprehensibility of systems plays a crucial role. While there is frequent discussion about abandonment of the Turing Test and focusing on more clearly specified tasks in well-defined domains, putting emphasis on making systems and their output comprehensible for humans offers an alternative approach to overcoming limitations of the original test, while still maintaining domain and task generality.

2.2 Predicate Invention

Predicate Invention, the automated introduction of auxiliary predicates, has been viewed as an important problem since the early days of ILP (e.g. [10,15,17]), though limited progress has been made in this topic recently [13]. Early approaches were based on the use of W-operators within the inverting resolution framework (e.g., [10,15]). However, the completeness of these approaches was never demonstrated, partly because of the lack of a declarative bias to delimit the hypothesis space. Failure to address these issues has, until recently, led to limited progress being made in this important topic and many well-known ILP systems such as ALEPH [16] and FOIL [14] have no predicate invention. In the recently introduced Meta-Interpretive Learning (MIL) framework [11,12], predicate invention is conducted via construction of substitutions for meta-rules applied by a meta-interpreter. The use of the meta-rules clarifies the declarative bias being employed. New predicate names are introduced as higher-order skolem constants, a finite number of which are added during every iterative deepening of the search.

3 Framework

3.1 General Setting

We assume sets of constants, predicate symbols and first-order variables are denoted $\mathcal{C}, \mathcal{P}, \mathcal{V}$. We assume definite clause programs to be defined in the usual way. Furthermore we assume a human as possessing background knowledge B expressed as a definite program. We now define the distinction between private and public predicate symbols.

Definition 1 [Public and private predicate symbols]. *We say that a predicate symbol $p \in \mathcal{P}$ found in definite program P is* public *with respect to a human population S in the case that p forms part of the background knowledge of each human $s \in S$. Otherwise p is* private.

Now we define Predicate Invention as follows.

Definition 2 [Predicate Invention]. *In the case background knowledge B of an ILP is extended to $B \cup H$, where H is a definite program we call predicate symbol $p \in \mathcal{P}$ an Invention iff p is defined in H but not in B.*

3.2 Comprehensibility

Next we provide our operational definition of comprehensibility.

Definition 3 [Comprehensibility, $C(S, P)$]. *The comprehensibility of a definition (or program) P with respect to a human population S is the mean accuracy with which a human s from population S after brief study and without further sight can use P to classify new material sampled randomly from the definition's domain.*

Note that this definition allows us to define comprehensibility in a way which allows its experimental determination given a set of human participants. However, in order to clarify the term "after brief study" we next define the notion of inspection time.

Definition 4 [Inspection time $T(S,P)$]. *The inspection time T of a definition (or program) P with respect to a human population S is the mean time that a human s from S spends studying P before applying P to new material.*

Since, in the previous subsection, we assume humans as having background knowledge which is equivalent to a definite program, we next define the idea of humans mapping privately defined predicate symbols to ones found in their own background knowledge.

Definition 5 [Predicate recognition $R(S,p)$]. *Predicate recognition R is the mean proportion of times that a human s from population S gives the correct public name to a predicate symbol p presented as a privately named definition q.*

For each of these mappings from privately defined predicate symbols to elements from the background knowledge we can now experimentally determine the required naming time.

Definition 6 [Naming time $N(S,p)$]. *For a predicate symbol p presented as a privately named definition q in definite program P the naming time N with respect to a human population S is the mean time that a human s from S spends studying P before giving a public name to p.*

Lastly we provide a simple definition of the textual complexity of a definite program.

Definition 7 [Textual complexity, $Sz(P)$]. *The textual complexity Sz of a definition of definite program P is the sum of the occurrences of predicate symbols, functions symbols and variables found in P.*

3.3   Experimental Hypotheses

We are now in a position to define and explain the motivations for the experimental hypotheses to be tested in Sect. 4. Below $C(S,P)$, $T(S,P)$, $R(S,p)$, $N(S,p)$, $Sz(P)$ are denoted by C, T, R, N and Sz respectively.

Hypothesis H1, $C \propto \frac{1}{T}$. This hypothesis relates to the idea of using inspection time as a proxy for incomprehension. That is, we might expect that humans take a long time to commit to an answer in the case they find the program hard to understand. As a proxy, inspection time is easier to measure than comprehension.

Hypothesis H2, $C \propto R$. This hypothesis is related to the idea that humans understand private predicate symbols, such as $p1/2$, generated during predicate invention, by mapping them to public ones in their own background knowledge.

Hypothesis H3, $C \propto \frac{1}{Sz}$. This hypothesis is motivated by the idea that a key property of predicate invention is its ability to compress a description by introducing new predicates which are used multiply within the definition. We are interested in whether the resultant compression of the description leads to increased comprehensibility.

Hypothesis H4, $R \propto \frac{1}{N}$**.** This hypothesis relates to the idea that if humans take a long time to recognise and publicly name a privately named predicate they are unlikely to correctly identify it. Analogous to H1, this allows naming time to be used as a proxy for recognition of an invented predicate.

In the next section we describe experiments which test these four hypotheses. Table 1 shows the mapping between the measurable properties defined in this section and the independent variables used in the experiment.

Table 1. Mapping defined properties from this section and independent variables in the experiment.

Defined property	Experimental variable
Comprehensibility C	Score
Inspection time T	Time
Recognition R	CorrectNaming
Naming Time N	NamingTime

4 Experiment

To investigate the hypotheses concerning comprehensibility and predicate invention, we conducted an experiment with human participants. In the following, we first present the material. Afterwards we present the independent and dependent variables and re-formulate the hypotheses with respect to these variables. Then we present the design and the results of the experiment. Finally, we relate the findings to the hypotheses of the framework.

4.1 Material

Material construction is based on the well-known family tree examples used to teach Prolog [18] and also used in the context of ILP [12]. Based on the *grandparent/2* predicate, three additional problems were defined: *grandfather/2* which is more specific than *grandparent/2*, *greatgrandparent/2* which needs the double amount of rules if defined without an additional (invented) predicate, that is, which has a high textual complexity, and the recursive predicate *ancestor/2* which has small textual but high cognitive complexity [5]. Instead of these meaningful names, target predicates are called *p/2*. Given facts are identical to the family tree presented in [12].[1] In the rule bodies, either public names (*mother, father*)—that is, names which relate to the well-known semantics of

---

[1] Please note that the relation *mother(matilda,bill)* needs to be changed to *mother(matilda,alice)* and the relation *father(jake,bill)* needs to be changed to *father(jake,alice)* to cover all cases necessary to invent the predicate *parent/2* in the context of different target predicates.

family relations—or private names ($q1/2$, $q2/2$) were used. Furthermore, programs were either presented with or without the inclusion of an additional (invented) predicate for *parent/2* which was named $p1/2$. The trees for the public and the private name space and the predicate definitions for the public name space are given in Fig. 2.

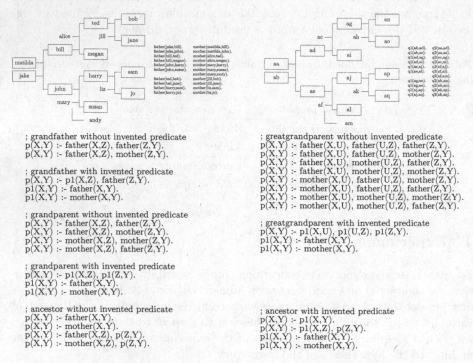

; grandfather without invented predicate
p(X,Y) :- father(X,Z), father(Z,Y).
p(X,Y) :- father(X,Z), mother(Z,Y).

; grandfather with invented predicate
p(X,Y) :- p1(X,Z), father(Z,Y).
p1(X,Y) :- father(X,Y).
p1(X,Y) :- mother(X,Y).

; grandparent without invented predicate
p(X,Y) :- father(X,Z), father(Z,Y).
p(X,Y) :- father(X,Z), mother(Z,Y).
p(X,Y) :- mother(X,Z), mother(Z,Y).
p(X,Y) :- mother(X,Z), father(Z,Y).

; grandparent with invented predicate
p(X,Y) :- p1(X,Z), p1(Z,Y).
p1(X,Y) :- father(X,Y).
p1(X,Y) :- mother(X,Y).

; ancestor without invented predicate
p(X,Y) :- father(X,Y).
p(X,Y) :- mother(X,Y).
p(X,Y) :- father(X,Z), p(Z,Y).
p(X,Y) :- mother(X,Z), p(Z,Y).

; greatgrandparent without invented predicate
p(X,Y) :- father(X,U), father(U,Z), father(Z,Y).
p(X,Y) :- father(X,U), father(U,Z), mother(Z,Y).
p(X,Y) :- father(X,U), mother(U,Z), father(Z,Y).
p(X,Y) :- father(X,U), mother(U,Z), mother(Z,Y).
p(X,Y) :- mother(X,U), father(U,Z), mother(Z,Y).
p(X,Y) :- mother(X,U), father(U,Z), father(Z,Y).
p(X,Y) :- mother(X,U), mother(U,Z), mother(Z,Y).
p(X,Y) :- mother(X,U), mother(U,Z), father(Z,Y).

; greatgrandparent with invented predicate
p(X,Y) :- p1(X,U), p1(U,Z), p1(Z,Y).
p1(X,Y) :- father(X,Y).
p1(X,Y) :- mother(X,Y).

; ancestor with invented predicate
p(X,Y) :- p1(X,Y).
p(X,Y) :- p1(X,Z), p(Z,Y).
p1(X,Y) :- father(X,Y).
p1(X,Y) :- mother(X,Y).

Fig. 2. Public tree (left), private tree (right) and the Prolog programs for *grandfather/2*, *grandparent/2*, *greatgrandparent/2*, and *ancestor/2* with and without use of an additional (invented) predicate *parent*. In the corresponding programs for the private name space, *father/2* is replaced by $q1/2$ and *mother/2* is replaced by $q2/2$.

In Sect. 3 we defined comprehensibility of a program as the accuracy with which a human can classify new material sampled from the domain. To assess comprehensibility, we defined seven questions for each of the four predicates (see Fig. 3). For five questions, it has to be determined whether a relation for two given objects is true. For two further questions, it has to be determined for which variable bindings the relation can be fulfilled. In addition, an open question was included, where a meaningful name had to be given to predicate $p/2$ for each of the four problems and—if applicable—also to the additional predicate $p1/2$.

To evaluate the material, we ran a pilot study (March 2016) at Imperial College London with 16 students of computer science with a strong background in programming, Prolog, and logic. The pilot study was conducted as a paper-and-pencil experiment where for each problem first the seven questions had to

– What is the result of p(bill,bob)?
  □ true   □ false   □ don't know
– What is the result of p(jake,harry)?
  □ true   □ false   □ don't know
– What is the result of p(bob,bill)?
  □ true   □ false   □ don't know
– What is the result of p(mary,jo)?
  □ true   □ false   □ don't know
– What is the result of p(john,sam)?
  □ true   □ false   □ don't know
– What is the result of p(X,bob)?
  □ false   □ X = bill   □ X = alice
  □ X = bill; alice   □ don't know
– What is the result of p(john,X)?
  □ false   □ X = sam   □ X = jo
  □ X = sam; jo   □ don't know

Fig. 3. Questions for the *grandparent/2* problem with public names.Questions for the *grandparent/2* problem with public names.

be answered and afterwards a meaningful name had to be given to the program. 13 out of the 16 students solved all questions correctly and most students were able to give the correct public names to all of the programs, regardless whether they had to work with the public or with the private names. Participants needed about 20 min for the four problems. Thus, the instructions and the material are understandable and coherent. A very interesting outcome of the study was that about a third of the students made notes on the questionnaires. Some of the notes showed that students first named the target predicates and the invented predicate (see Fig. 1) and then answered the questions. That is, students gave a meaningful name without being instructed to do so and one can assume that they used this strategy because it made answering the questions easier.

4.2   Variables and Empirical Hypotheses

To assess the influence of meaningful names and of predicate invention on comprehensibility, we introduced the following three independent variables:

NameSpace: The name space in which context the problems is presented is either **public** or **private** as shown in Fig. 2.

PredicateInvention: The problems are given either **with** or **without** an additional (invented) predicate *p1/2* which represents the *parent/2* relation.

NamingInstruction: The open question to give a meaningful name to predicate *p/2* is either given **before** or **after** the seven questions given in Fig. 3 had to be answered.

The variation of the independent variables results in a $2 \times 2 \times 2$ factor design which was realised between-participants for factors NameSpace and NamingInstruction and within-participants for factor PredicateInvention. Problem presentation with PredicateInvention was either given for the first and the third or the second and the fourth problem.

The **textual complexity** varies over problems and in dependence of the introduction of the additional predicate *p1/2*. The textually most complex program is *greatgrandparent/2* without the use of *p1/2*. The least complex program is *grandfather/2* without the use of *p1/2* as can be seen in Fig. 2.

The following dependent variables were assessed:

Score: For each problem, the score is calculated as the sum of correctly answered questions (see Fig. 3). That is, score has minimal value 0 and maximal value 7 for each problem.

Time: The time to inspect a problem is measured from presenting the problem until answering the seven questions.

CorrectNaming: The correctness of the given public name for a predicate definition *p/2* was judged by two raters. In addition, it was discriminated between clearly incorrect answers and responses where participants wrote nothing or stated that they do not know the correct meaning.

NamingTime: The time for naming is measured from presenting the question until indication that the question is answered by going to the next page. For condition PredicateInvention/with both *p/2* and *p1/2* had to be named.

Given the independent and dependent variables, hypotheses can now be formulated with respect to these variables:

H1: Score is inverse proportional to Time, that is, participants who comprehend a program, give more correct answers in less time than such participants who do not comprehend the program.

H2: CorrectNaming is proportional to Score, that is, participants who can give the intended public—that is, meaningful—name to a program have higher scores than participants who do not get the meaning of the program.

H3: Score is inverse proportional to textual complexity, that is, for problem *greatgrandparent/2* the differences of score should be greatest between the PredicateInvention/with and PredicateInvention/without condition because here the difference in textual complexity is highest.

H4: CorrectNaming is inverse proportional to NamingTime, that is, if participants need a long time to come up with a meaningful name for a program, they probably will get it wrong.

4.3   Participants and Procedure

The experiment was conducted in April 2016 with cognitive science students of the University of Osnabrueck. All students had passed at least one previous one-semester course on Prolog programming and all have a background in logic. That is, their background in Prolog is less strong than for the Imperial College sample but they are no novices. From the originally 87 participants, three did not finish the experiment and six students were excluded because they answered "don't know" for more than 50% of the questions. All analyses were done with the remaining 78 participants (43 male, 35 female; mean age 23.55 years, sd = 2.47).[2]

The experiment was realised with the soscisurvey.de system and was conducted online during class. After a general introduction, students worked through an example problem ("sibling") to get acquainted with the domain—that is either the family tree or the abstract tree shown in Fig. 2—and with the types of questions they needed to answer. Afterwards, the four test problems were presented in one of the eight experimental conditions. For each problem, on the first page the facts and the tree and the predicate definition was presented. On the next page, this information was given again together with the first question or the naming instruction. If the "next"-button was pressed, it was not possible to go back to a previous page.

Working through the problems was self-paced. The four problems were presented in the sequence *grandfather/2*, *grandparent/2*, *greatgrandparent/2*, *ancestor/2* for all participants. That is, we cannot control for sequence effects, such as performance gain due to getting acquainted with the style of the problems and questions or performance loss due to decrease in motivation or fatigue. However, since problem type is not used as an experimental condition, possible sequence effects do not affect statistical analyses of the effects of the independent variables introduced above.

4.4   Results

Scores and Times. When considering time for question answering and naming together, participants needed about 5 minutes for the first problem and got faster over the problems. One reason for this speed-up effect might be, that participants needed less time to inspect the tree or the facts for later problems. There is no speed-accuracy trade-off, that is, there is no systematic relation between (low) number of correct answers and (low) solution time for question answering. In the following, time is given in seconds and for statistical analyses time was logarithmically transformed.

Giving Meaningful Names. In the public name condition, the names the participants gave to the programs were typically the standard names, sometimes their inverse, such as "grandchildren", "child of child", or "parent of parent" for the *grandparent/2* problem. In the condition with private names, the standard names describing family relations were also used by most participants, however, some participants gave more abstract descriptions, such as "X and Y are connected via an internode" for *grandparent/2*. Among the incorrect answers for the *grandparent/2* problem often were over-specific interpretations such as "grandson" or "grandfather". The same was the case for *greatgrandparent/2* with incorrect answers such as "greatgrandson". Some participants restricted the description to the given tree, for example, "parent of parent with 2 children"

---

[2] A comprehensive description of all analyses and results can be found at http://www.cogsys.wiai.uni-bamberg.de/publications/comprAnalysesDoc.pdf.

for *grandparent/2*. Incorrect answers for the *ancestor/2* problem typically were overly general, such as "related".

Impact of NameSpace, PredicateInvention, and NamingInstruction on Score and Time. An overview of the impact of all factors on score is given in Fig. 4. There it can be seen that NameSpace/public results in higher scores for all four problems. The effects of PredicateInvention and NamingInstruction are less obvious. It is not the case that having to think about the meaning of a predicate before question answering has a general positive effect on Score. PredicateInvention is helpful for some problems, for others not. We will give a closer look on the effect of PredicateInvention for the textually most complex problem *greatgrandparent/2* below (H3). Statistical analyses were done with general linear models with NameSpace, PredicateInvention, and NamingInstruction as predictor variables and Score as criterion variable. Predictor variables were dummy coded as contrasts. The effect of NameSpace/public is significant for *grandfather/2* ($b = 1.55$, $p = 0.03$) and marginally significant for *greatgrandparent/2* ($b = 1.12$, $p = 0.069$). In addition, for *grandfather/2* the interaction of NameSpace and PredicateInvention is significant ($b = -2.52$, $p = 0.017$).

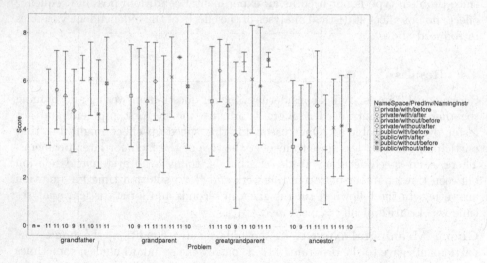

Fig. 4. Scores distributed over NameSpace, PredicateInvention, and NamingInstruction (arithmetic means and standard deviations are given; for significant differences, see text).

Inverse proportional relation between Score and Time (H1). There is a significant negative Pearsons product-moment correlation between Time and Score over all problems ($r = -.38$, $p \leq 0.001$).

Effect of CorrectNaming on Score (H2). To assess the impact of being able to give a meaningful name to a problem (CorrectNaming) on comprehensibility (Score), answers were classified as "correct", "incorrect" and "no answer" which covers answers where participants either did not answer or explicitly stated that they do not know the answer. Participants who were able to give meaningful names to the programs answered significantly more questions correctly. Statistical analyses were again performed with general linear models with dummy coding (contrast) for the predictor variable CorrectNaming. The results are given in Table 2.

Table 2. Means and standard deviations of Score in dependence of CorrectNaming, where "no answer" covers answers where participants either did not answer or explicitly stated that they do not know the answer. Results for linear models are given as b-estimates and p-values for the contrast between correct and incorrect naming.

	Correct	Incorrect	No answer	Test
Grandfather	n = 28	n = 46	n = 4	
Score	Mean 6.68 (sd = 0.61)	5.15 (1.81)	4.75 (1.71)	$-1.53, p < 0.001$
Grandparent	50	23	5	
Score	6.56 (1.23)	5.04 (2.12)	3.4 (1.82)	$-1.52, p < 0.001$
Greatgrandparent	54	18	6	
Score	6.76 (0.66)	5.78 (1.66)	3 (1.67)	$-1, p < 0.001$
Ancestor	32	39	7	
Score	5.75 (1.44)	3.08 (1.8)	2.86 (1.57)	$-2.67, p < 0.001$

Impact of textual complexity on the effect of PredicateInvention on Score (H3). For the *greatgrandparent/2* problem, there is a marginally significant effect of PredicateInvention for NameSpace/private and NamingInstruction/after with a higher score for the PredicateInvention/with condition ($b = -1.59, p = 0.09$).

Relation of CorrectNaming and NamingTime (H4). Participants who give a correct meaningful name to a problem do need less time to do so than participants who end up giving an incorrect name for all problems except *ancestor/2*. This relation is given in Fig. 5 accumulated over all factors per problem. Statistical analyses were done separately for conditions PredicateInvention/with and PredicateInvention/without because in the first case two names—for target predicate $p/2$ and for the additional predicate $p1/2$—had to be given. Differences between correct and incorrect are significant for *grandfather/2* in the condition PredicateInvention/without ($b = 0.31, p = 0.007$) and marginally significant for *grandparent/2* in the condition PredicateInvention/with ($b = 0.2, p = 0.084$). For *ancestor/2* in the condition PredicateInvention/with there is a significant difference between correct naming and "no answer" ($b = -0.49, p = 0.039$).

4.5   Interpretation and Discussion

Results show that presenting programs in relation to a public name space facilitates comprehension. Contrary to our expectations, being instructed to first think about a meaningful name for a program before answering questions in general does not facilitate generation of answers. We would have expected that having a (denotational) semantic interpretation for a predicate supports working on classification and variable bindings of new material from a given domain because mental evaluation of a program can be—at least partially—avoided. Furthermore, as expected, the use of additional (invented) predicates does not facilitate program comprehension in general but only under specific conditions which are discussed below (H3).

Results concerning our hypotheses are summarised in Table 3. Hypothesis H1 is confirmed by our empirical data: if a person comprehends a program, she or he can come up with correct answers in short time. Hypothesis H2 is also confirmed: participants who can give a meaningful name to a program give more correct answers than participants who give incorrect answers or state that they do not know the answer. In addition, participants who give a correct name give answers faster. As hypothesis H3 we assumed that predicate invention supports comprehensibility if it reduces the

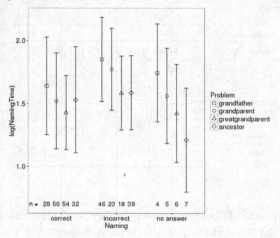

Fig. 5. Relation between time needed for giving a meaningful name and correctness of naming, where "no answer" covers answers where participants either did not answer or explicitly stated that they do not know the answer (averaged over PredicateInvention with/without).

textual complexity of a program. For the four problems we investigated, the reduction in complexity is greatest for *greatgrandparent/2*. Here we get a partial confirmation: predicate invention results in more correct answers for the private name space and if the instruction for naming was given after question answering. This experimental condition is the most challenging, because comprehensibility is not supported by public names and because participants were not encouraged to think about the meaning of the presented predicate before they had to answer questions about it.

Finally, we assumed that persons who have problems to come up with a meaningful name for a predicate spend a longer amount of time to come up with an (incorrect or no) answer (H4). Results show that this is the case— with the exception of the *ancestor/2* problem. However, the differences are only significant under specific conditions. The observation that long answering time can indicate a problem with comprehensibility could be exploited for the design of the interaction of a person with an ILP system: if a person does not come up quickly with a name for a predicate, the system could offer examples of the predicates behaviour. For example, for the *ancestor/2* problem, pairs for which this predicate is true could be highlighted in the given tree.

It can be assumed that the empirical results depend on the level of expertise of the participants. As we saw, the highly experienced sample of students of Imperial College did not profit from public name space or from the use of invented predicates. They answered most questions correctly under all conditions. In contrast, for the moderately experienced sample of students from University of Osnabrueck, presenting predicates in relation to a public name space and under some conditions with invented predicates resulted in better comprehensibility. For a sample of Prolog novices, the experimental variations might result in stronger or different effects.

Table 3. Hypotheses concerning comprehensibility, meaningful names, and predicate invention.

Hypothesis		Confirmation
H1	Comprehensibility manifests itself in high scores and fast solution times	Confirmed
H2	Comprehensibility means to be able to give a meaningful name to a program	Confirmed
H3	Predicate invention helps comprehensibility if it reduces textual complexity of the program	Partially
H4	If coming up with a meaningful name needs a long time, it will probably be the false concept	Partially

5   Conclusions and Further Work

This paper is, to our knowledge, the first paper in the literature which provides an operational definition of the comprehensibility of a logic program. The definition is used within the experiments in Sect. 4 to identify factors which affect comprehension. These factors include the time required to inspect the program, the accuracy with which a participant can recognise a predicate to be equivalent to one already known and the textual complexity of the program.

As expected, the four problems presented in the experiment differ with respect to comprehensibility. The problem most participants had difficulty with is the recursive *ancestor/2*. For this problem less than half of the participants (32) gave the correct meaningful name and for this problem participants have the lowest scores. However, since this problem was positioned last in the sequence, the results might also be due to loss of motivation or exhaustion. Astonishingly, *ancestor/2* is also the only of the four problems where participants reached the highest score in the private naming condition without predicate invention (cf. Fig. 4). We plan a follow-up experiment where problem sequences are varied to determine whether this is a systematic effect.

The kinship predicates presented to human participants in our experiments are all ones which could be expected to be equivalent to ones already known to the participant. In further work we hope also to study the effects of human users being presented with definitions of predicates which are novel for the user.

In closing we believe the operational definition of comprehensibility has enormous potential to both clarify one of the central concepts of AI research, as well as to provide a bridge to the study of factors affecting the design of AI systems which improve human understanding.

References

1. Cropper, A., Muggleton, S.H.: Learning efficient logical robot strategies involving composable objects. In: Proceedings of the 24th International Joint Conference Artificial Intelligence (IJCAI 2015), pp. 3423–3429 (2015)
2. Cropper, A., Muggleton, S.H.: Learning higher-order logic programs through abstraction and invention. In: Proceedings of the 25th International Joint Conference Artificial Intelligence (IJCAI 2016), pp. 1418–1424 (2016)
3. Forbus, K.D.: Software social organisms: implications for measuring AI progress. AI Mag. **37**(1), 85–90 (2016)
4. Freitas, A.A.: Comprehensible classification models: a position paper. SIGKDD Explor. Newsl. **15**(1), 1–10 (2014)
5. Kahney, H.: What do novice programmers know about recursion? In: Soloway, E., Spohrer, J.C. (eds.) Studying the Novice Programmer, pp. 209–228. Lawrence Erlbaum (1989)
6. Letham, B., Rudin, C., McCormick, T.H., Madigan, D.: Interpretable classifiers using rules and Bayesian analysis: building a better stroke prediction model. Ann. Appl. Stat. **9**(3), 1350–1371 (2015)
7. Lin, D., Dechter, E., Ellis, K., Tenenbaum, J.B., Muggleton, S.H.: Bias reformulation for one-shot function induction. In: Proceedings of the 23rd European Conference on Artificial Intelligence (ECAI 2014), pp. 525–530. IOS Press (2014)
8. Michie, D.: Machine learning in the next five years. In: Proceedings of the Third European Working Session on Learning, pp. 107–122. Pitman (1988)
9. Mozina, M., Zabkar, J., Bratko, I.: Argument based machine learning. Artif. Intell. **171**(10–15), 922–937 (2007)
10. Muggleton, S.H., Buntine, W.: Machine invention of first-order predicates by inverting resolution. In: Proceedings of the 5th International Conference on Machine Learning, pp. 339–352. Kaufmann (1988)
11. Muggleton, S.H., Lin, D., Pahlavi, N., Tamaddoni-Nezhad, A.: Meta-interpretive learning: application to grammatical inference. Mach. Learn. **94**, 25–49 (2014)
12. Muggleton, S.H., Lin, D., Tamaddoni-Nezhad, A.: Meta-interpretive learning of higher-order dyadic datalog: predicate invention revisited. Mach. Learn. **100**(1), 49–73 (2015)
13. Muggleton, S.H., De Raedt, L., Poole, D., Bratko, I., Flach, P., Inoue, K.: ILP turns 20: biography and future challenges. Mach. Learn. **86**(1), 3–23 (2011)
14. Quinlan, J.R.: Learning logical definitions from relations. Mach. Learn. **5**, 239–266 (1990)

15. Rouveirol, C., Puget, J.-F.: A simple and general solution for inverting resolution. In: Proceedings of the fourth European Working Session on Learning (EWSL-1989), pp. 201–210. Pitman (1989)
16. Srinivasan, A.: The ALEPH manual. Machine Learning at the Computing Laboratory, Oxford University (2001)
17. Stahl, I.: Constructive induction in inductive logic programming: an overview. Technical report, Fakultät Informatik, Universität Stuttgart (1992)
18. Sterling, L., Shapiro, E.Y.: The Art of Prolog: Advanced Programming Techniques. MIT Press, Cambridge (1994)
19. Turing, A.M.: Computing machinery and intelligence. Mind 59(236), 433–460 (1950)

Distributional Learning of Regular Formal Graph System of Bounded Degree

Takayoshi Shoudai[1]([⊠]), Satoshi Matsumoto[2], and Yusuke Suzuki[3]

[1] Faculty of International Studies, Kyushu International University,
Kitakyushu, Japan
shoudai@isb.kiu.ac.jp
[2] Faculty of Science, Tokai University, Hiratsuka, Japan
[3] Graduate School of Information Sciences,
Hiroshima City University, Hiroshima, Japan

Abstract. In this paper, we describe how distributional learning techniques can be applied to formal graph system (FGS) languages. An FGS is a logic program that deals with term graphs instead of the terms of first-order predicate logic. We show that the regular FGS languages of bounded degree with the 1-finite context property (1-FCP) and bounded treewidth property can be learned from positive data and membership queries.

1 Introduction

In the field of algorithmic learning theory, many models and algorithmic techniques for learning from examples have been developed. Distributional learning was first proposed by Clark and Eyraud [3] to learn a subclass of context-free grammars efficiently. Recently, distributional learning techniques have been developed for learning of various subclasses of context-free grammars [11]. These techniques were extended to languages that have more complex structures [7]. Yoshinaka [12] introduced distributional properties on grammars and showed that grammars with distributional properties are learnable with standard distributional learning techniques if the grammars satisfy certain conditions, e.g. polynomial time decomposability of objects into contexts and substructures.

Graph grammar has been developed as an extension to graphs from strings of grammatical forms. Graph grammar has been applied to a wide range of fields including pattern recognition and image analysis. Uchida et al. [10] introduced a framework called formal graph system (FGS) as a graph grammar. An FGS is a logic program that deals with term graphs, which can be considered to be types of hypergraphs, instead of the terms of first-order predicate logic.

For the learning of graph grammar, Okada et al. [9] showed that some classes of graph pattern languages are learned from a minimally adequate teacher (MAT) in polynomial time. Hara and Shoudai [6] proposed an algorithm for

---

T. Shoudai—This work was partially supported by JSPS KAKENHI (26280087, 15K00313) and MEXT KAKENHI (24106010).

J. Cussens and A. Russo (Eds.): ILP 2016, LNAI 10326, pp. 68–80, 2017.
DOI: 10.1007/978-3-319-63342-8_6

learning the class of c-deterministic regular FGS languages in the framework of MAT learning. There have been other studies on graph grammars from the viewpoint of application, but discussions on computational learning of graph grammars are not yet sufficient. In this paper, we show that the regular FGS languages of bounded degree with the 1-finite context property (1-FCP) [2] and bounded treewidth property can be learned from positive data and membership queries with current distributional learning techniques [11].

2   Preliminaries

For a set or a list S, $|S|$ denotes the number of all elements that are contained in S. For a set S, S^* denotes the set of all finite lists consisting of elements in S. For a list S and an integer i $(1 \leq i \leq |S|)$, $S[i]$ denotes the i-th member of S. Let Σ and Λ be finite alphabets. Let X be an infinite alphabet, whose elements are called *variables*. We assume that each symbol $x \in X$ has a nonnegative integer $rank(x)$, $\Sigma \cap X = \emptyset$ and $\Lambda \cap X = \emptyset$.

Definition 1 (Term graph). *A term graph $g = (V, E, \varphi, \psi, H, \lambda, ports)$ is defined as follows:*

1. *(V, E) is a vertex- and edge-labeled (directed or undirected) graph,*
2. *$\varphi : V \rightarrow \Sigma$ and $\psi : E \rightarrow \Lambda$ are vertex- and edge-labeling functions,*
3. *H is a finite multiset of hyperedges that are elements of 2^V,*
4. *$\lambda : H \rightarrow X$ is a variable-labeling function, and*
5. *ports : $H \rightarrow V^*$ is a mapping s.t. for every $h \in H$, $ports(h)$ is a list of $rank(\lambda(h))$ distinct vertices in V. These vertices are called the ports of h.*

We give an example of term graphs in Fig. 1. A hyperedge is drawn as a box with lines to its ports. The order of the ports is indicated by digits at these lines.

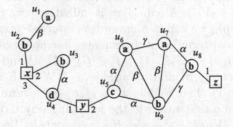

Fig. 1. A term graph $g = (V, E, \varphi, \psi, H, \lambda, ports)$ on $\Sigma = \{\mathbf{a}, \mathbf{b}, \mathbf{c}, \mathbf{d}\}$ and $\Lambda = \{\alpha, \beta, \gamma\}$: $H = \{\{u_2, u_3, u_4\}, \{u_4, u_5\}, \{u_8\}\}, \lambda(\{u_2, u_3, u_4\}) = x, \lambda(\{u_4, u_5\}) = y, \lambda(\{u_8\}) = z, ports(\{u_2, u_3, u_4\}) = (u_2, u_3, u_4), ports(\{u_4, u_5\}) = (u_4, u_5), ports(\{u_8\}) = (u_8)$.

For a term graph g, its 7-tuple is denoted by $(V_g, E_g, \varphi_g, \psi_g, H_g, \lambda_g, ports_g)$. A term graph g is called *ground* if $H_g = \emptyset$ and both λ_g and $ports_g$ are empty functions $\emptyset$. We define the size of a term graph g, denoted by $|g|$, as $|V_g| + |E_g| + |H_g|$. A term graph g is a *star term graph* if $E_g = \emptyset$ and $H_g = \{V_g\}$. For a star term graph g, h_g denotes the unique hyperedge of g.

Definition 2 (Treewidth [5]**).** *A tree decomposition of a term graph* $g = (V, E, \varphi, \psi, H, \lambda, ports)$ *is a rooted tree* $\mathcal{T} = (\mathcal{I}, \mathcal{F})$ *whose vertices* $i \in \mathcal{I}$ *are associated with* $V_i \subseteq V$, $E_i \subseteq E$ *and* $H_i \subseteq H$ *that satisfy the following conditions:*

1. *For each* $v \in V$, *there is a vertex* $i \in \mathcal{I}$ *such that* $v \in V_i$.
2. *For each* $e = (u, v) \in E$, *there is exactly one vertex* $i \in \mathcal{I}$ *such that* $u, v \in V_i$ *and* $e \in E_i$.
3. *For each* $h = \{v_1, \ldots, v_m\} \in H$, *there is exactly one vertex* $i \in \mathcal{I}$ *such that* $v_1, \ldots, v_m \in V_i$ *and* $h \in H_i$.
4. *For each* $v \in V$, *the subtree of* $\mathcal{T}$ *induced by* $\{i \in \mathcal{I} \mid v \in V_i\}$ *is connected.*

The *width* of $\mathcal{T}$ is defined as $\max_{i \in \mathcal{I}} |V_i| - 1$. The *treewidth* of g is the minimum width of any tree decomposition $\mathcal{T} = (\mathcal{I}, \mathcal{F})$ of g.

Two term graphs f and g are said to be *isomorphic*, if there is a bijection π from V_f to V_g, such that (1) $(u, v) \in E_f$ if and only if $(\pi(u), \pi(v)) \in E_g$, (2) $\varphi_f(u) = \varphi_g(\pi(u))$ for each vertex $u \in V_f$ and $\psi_f(u, v) = \psi_g(\pi(u), \pi(v))$ for each edge $(u, v) \in E_f$, (3) $\{v_1, \ldots, v_\ell\} \in H_f$ if and only if $\{\pi(v_1), \ldots, \pi(v_\ell)\} \in H_g$, (4) $\lambda_f(\{v_1, \ldots, v_\ell\}) = \lambda_f(\{u_1, \ldots, u_\ell\})$ if and only if $\lambda_g(\{\pi(v_1), \ldots, \pi(v_\ell)\}) = \lambda_g(\{\pi(u_1), \ldots, \pi(u_\ell)\})$ for each hyperedge $\{v_1, \ldots, v_\ell\}, \{u_1, \ldots, u_\ell\} \in H_f$, and (5) $ports_f(\{v_1, \ldots, v_\ell\}) = ports_g(\{\pi(v_1), \ldots, \pi(v_\ell)\})$ for each $\{v_1, \ldots, v_\ell\} \in H_f$. A bijection π satisfying (1)–(5) is called an *isomorphism* from f to g.

Let d and w be nonnegative integers. The degree of a vertex v is defined as $|\{e \in E_g \mid v \in e\}| + |\{h \in H_g \mid v \in h\}|$. $\mathcal{G}(\Sigma, \Lambda, X)$ (resp. $\mathcal{G}_{d,w}(\Sigma, \Lambda, X)$) denotes the set of all term graphs (resp. all term graphs of maximum degree d and treewidth w) over $\langle \Sigma, \Lambda, X \rangle$. Moreover, $\mathcal{G}(\Sigma, \Lambda)$ (resp. $\mathcal{G}_{d,w}(\Sigma, \Lambda)$) denotes the set of all ground term graphs (resp. all ground term graphs of maximum degree d and treewidth w).

Let f be a term graph in $\mathcal{G}(\Sigma, \Lambda, X)$ and σ an ordered list of ℓ distinct vertices in V_f ($0 \leq \ell \leq |V_f|$). A pair $[f, \sigma]$ is called a *term graph fragment*. If f is a ground term graph, we call it a *ground term graph fragment*. Let $\mathcal{F}(\Sigma, \Lambda)$ be the set of all ground term graph fragments. For nonnegative integers d and w, $\mathcal{F}_{d,w}(\Sigma, \Lambda) = \{[f, \sigma] \in \mathcal{F}(\Sigma, \Lambda) \mid f \in \mathcal{G}_{d,w}(\Sigma, \Lambda)$ and $|\sigma| \leq d + 1\}$. For a term graph fragment $[f, \sigma]$ and a variable $x \in X$ with $rank(x) = |\sigma|$. Let $\sigma = (v_1, \ldots, v_\ell)$ ($\ell \geq 1$). The *binding* $x := [f, \sigma]$ for a term graph g is defined to be an operation on g that works in the following way: For each $h \in H_g$ with $\lambda_g(h) = x$, let $f' = (V_{f'}, E_{f'}, \varphi_{f'}, \psi_{f'}, H_{f'}, \lambda_{f'}, ports_{f'})$ be a copy of f. For a vertex $v \in V_f$, we denote the corresponding copy vertex of f' by v'. We attach f' to g by removing the hyperedge h from H_g and by identifying the ports $u_1, \ldots, u_\ell$ of h in g with $v'_1, \ldots, v'_\ell$ in f', respectively. We set the new vertex-label of u_i to be the original vertex-label of u_i, i.e., $\varphi_g(u_i)$. A *substitution* θ is a finite set of bindings $\{x_1 := [f_1, \sigma_1], \ldots, x_n := [f_n, \sigma_n]\}$, where x_i's are mutually distinct variables in X and each f_i has no hyperedge labeled with a variable in $\{x_1, \ldots, x_n\}$. We give an example of term graphs and substitutions in Fig. 2.

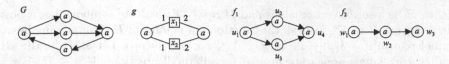

Fig. 2. A graph G can be obtained from g by applying a substitution $\theta = \{x_1 := [f_1, (u_1, u_4)], x_2 := [f_2, (w_3, w_1)]\}$, i.e., $g\theta$ is isomorphic to G.

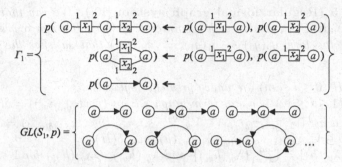

Fig. 3. A formal graph system $S_1 = (\Sigma_1, \Lambda_1, X_1, \Pi_1, \Gamma_1)$, where $\Sigma_1 = \{a\}, \Lambda_1 = \{\epsilon\}, X_1 = \{x_1, x_2, \ldots\}, \Pi_1 = \{p\}$, and its FGS language $GL(S_1, p)$.

Definition 3 (Formal graph system [10]). *Let $g_1, \ldots, g_n \in \mathcal{G}(\Sigma, \Lambda, X)$ ($n \geq 1$). Let Π_n be a finite set of n-ary predicate symbols. Let $\Pi = \bigcup_{i \geq 0} \Pi_i$. For $p \in \Pi_n$, we say that $p(g_1, \ldots, g_n)$ is an atom. Let $A, B_1, \ldots, B_m$ be atoms ($m \geq 0$) consisting of term graphs in $\mathcal{G}(\Sigma, \Lambda, X)$ and predicate symbols in Π. A graph rewriting rule over $\langle \Sigma, \Lambda, X, \Pi \rangle$ is a clause of the form $A \leftarrow B_1, \ldots, B_m$. For a clause $A \leftarrow B_1, \ldots, B_m$, the atom A is called the head and the right hand side of the arrow $B_1, \ldots, B_m$ is called the body of the rule. Let Γ be a finite set of graph rewriting rules over $\langle \Sigma, \Lambda, X, \Pi \rangle$. A formal graph system (abbreviated to FGS) is the 5-tuple $S = (\Sigma, \Lambda, X, \Pi, \Gamma)$.*

For a substitution θ and an atom $p(g_1, \ldots, g_n)$, we define $p(g_1, \ldots, g_n)\theta$ to be $p(g_1\theta, \ldots, g_n\theta)$. For a graph rewriting rule $A \leftarrow B_1, \ldots, B_m$, we also define $(A \leftarrow B_1, \ldots, B_m)\theta$ to be $A\theta \leftarrow B_1\theta, \ldots, B_m\theta$.

Definition 4. *Let $S = (\Sigma, \Lambda, X, \Pi, \Gamma)$ be an FGS. For a clause C, relation $\Gamma \vdash C$ is defined recursively in the following way:*

1. *If $C \in \Gamma$, then $\Gamma \vdash C$ holds.*
2. *If $\Gamma \vdash C$, then $\Gamma \vdash C\theta$ for an arbitrary substitution θ.*
3. *If $\Gamma \vdash A \leftarrow B_1, \ldots, B_n$ and for some i ($1 \leq i \leq n$), $\Gamma \vdash B_i \leftarrow C_1, \ldots, C_m$, then $\Gamma \vdash A \leftarrow B_1, \ldots, B_{i-1}, C_1, \ldots, C_m, B_{i+1}, \ldots, B_n$ holds.*

For an FGS $S = (\Sigma, \Lambda, X, \Pi, \Gamma)$ and a unary predicate symbol p, we define the *graph language* of (S, p) as $GL(S, p) = \{g \in \mathcal{G}(\Sigma, \Lambda) \mid \Gamma \vdash p(g) \leftarrow\}$. We say that a graph language $L \subseteq \mathcal{G}(\Sigma, \Lambda)$ is *definable by an FGS* or an *FGS language* if such a pair (Γ, p) exists. In Fig. 3, we give an example of the FGSs and its FGS language.

Let Σ be a set of vertex labels and Π a set of predicate symbols. Let δ be a function from Π to Σ^*. We call the function δ a *pointer* of Π if for any predicate symbol p, $\delta(p)[i] \neq \delta(p)[j]$ for all i, j $(1 \leq i < j \leq |\delta(p)|)$. Let $\delta(\Pi)$ be the set of all vertex labels appearing in $\delta(p)$ for all predicate symbols $p \in \Pi$. For a term graph g and a list of vertices $\sigma = (v_1, \ldots, v_\ell) \in V_g^\ell$ $(\ell \geq 1)$, $\varphi_g(\sigma)$ denotes $(\varphi_g(v_1), \ldots, \varphi_g(v_\ell))$.

Definition 5 (Regular formal graph system [10]**).** *We say that an FGS $S = (\Sigma, \Lambda, X, \Pi, \Gamma)$ is regular with a pointer δ of Π if all graph rewriting rules in Γ are of the form $q_0(g_0) \leftarrow q_1(g_1), \ldots, q_m(g_m)$ that satisfies the following conditions:*

1. *All $q_i \in \Pi$ $(0 \leq i \leq m)$ are unary predicate symbols.*
2. *Each g_i $(1 \leq i \leq m)$ is a star term graph s.t. $\varphi_{g_i}(ports_{g_i}(h_{g_i})) = \delta(q_i)$.*
3. *There is a list $(v_1, \ldots, v_{|\delta(q_0)|}) \in V_{g_0}^{|\delta(q_0)|}$ s.t. $\varphi_{g_0}(v_1, \ldots, v_{|\delta(q_0)|}) = \delta(q_0)$ and for any $u \in V_{g_0} \backslash \{v_1, \ldots, v_{|\delta(q_0)|}\}$, $\varphi_{g_0}(u) \in \Sigma \backslash \delta(\Pi)$.*
4. *$\bigcup_{h \in H_{g_0}} \{\lambda_{g_0}(h)\} = \bigcup_{i=1}^m \{\lambda_{g_i}(h_{g_i})\}$ and $\lambda_{g_i}(h_{g_i}) \neq \lambda_{g_j}(h_{g_j})$ for $1 \leq i < j \leq m$.*
5. *For every $h_1, h_2 \in H_{g_0}$, $h_1 \neq h_2$ if and only if $\lambda_{g_0}(h_1) \neq \lambda_{g_0}(h_2)$.*

A regular FGS $S = (\Sigma, \Lambda, X, \Pi, \Gamma)$ with a pointer δ is denoted by (S, δ) or $((\Sigma, \Lambda, X, \Pi, \Gamma), \delta)$. Below we call a regular FGS with a pointer a regular FGS.

Let (S, δ) be a regular FGS and p a unary predicate symbol in Π. We define the *graph language* of (S, δ, p) as $GL(S, \delta, p) = \{g \in \mathcal{G}(\Sigma, \Lambda) \mid \Gamma \vdash p(g) \leftarrow\}$. We say that a graph language $L \subseteq \mathcal{G}(\Sigma, \Lambda)$ is *definable by a regular FGS* or a *regular FGS language* if a triplet (S, δ, p) exists such that $L = GL(S, \delta, p)$. In Fig. 4, we give an example of the regular FGSs and its regular FGS language.

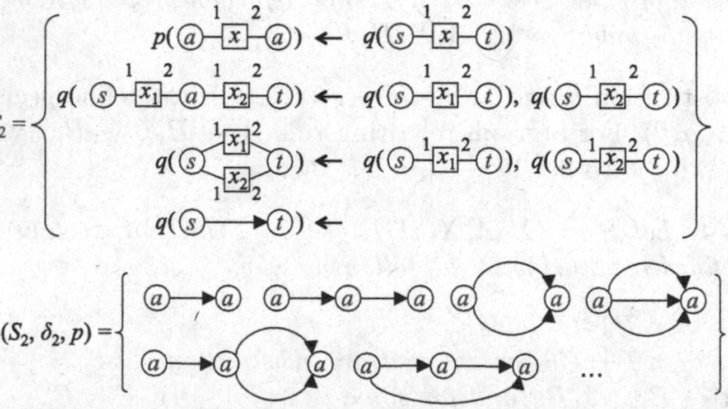

Fig. 4. A regular formal graph system $(S_2, \delta_2) = ((\Sigma_2, \Lambda_2, X_2, \Pi_2, \Gamma_2), \delta_2)$, where $\Sigma_2 = \{a, s, t\}, \Lambda_2 = \{\epsilon\}, X_2 = \{x_1, x_2, \ldots\}, \Pi_2 = \{p, q\}, \delta_2(p) = (), \delta_2(q) = (s, t)$, and its FGS language $GL(S_2, \delta_2, p)$, which is equivalent to the set of all two terminal series parallel graphs (TTSP graphs). Every TTSP graph has treewidth at most 2.

Definition 6 (Chomsky normal form). *Let f_0 be a ground term graph of one vertex or two vertices with one edge, f_1 a term graph with two hyperedges and no edge, and f_2, f_3 star term graphs. Let p_0, p_1, p_2, p_3 be unary predicate symbols. A regular FGS (S, δ) is in Chomsky normal form if every graph rewriting rule of S is of the form.*

- *Terminal rule: $p_0(f_0) \leftarrow$,*
- *Unary rule: $p_1(f_1) \leftarrow p_2(f_2)$.*
- *Binary rule: $p_1(f_1) \leftarrow p_2(f_2), \; p_3(f_3)$.*

The regular FGS in Figs. 3 and 4 is written in Chomsky normal form.

We say that a term graph g is connected if for any two vertices u and v of g, there is a sequence of vertices $v_0(= u), v_1, \ldots, v_m(= v)$ for an integer m such that for all i ($0 \leq i \leq m - 1$), v_i and v_{i+1} are contained in the same edge or hyperedge. In this paper, we assume that all term graphs are connected.

Graph grammar has been defined in various ways. One of the famous context-free graph grammars is a hyperedge replacement grammar (HRG) [4]. Uchida et al. [10] showed that a class of graphs is generated by an HRG if and only if it is defined by a regular FGS. This result shows that regular FGSs can generate interesting graph classes including trees, two-terminal series parallel graphs (in Fig. 4), and so on.

In the research of HRGs, Lautemann [8] gave some conditions on either grammar or input graphs whose parsing can be done in polynomial time. A parsing algorithm due to Lautemann is known to be polynomial time for graphs that are connected and of bounded degree. As a more precise characterization of the algorithm's complexity, Chiang et al. [1] showed that the parsing algorithm runs in polynomial time if the maximum degree and treewidth of graphs in an HRG are bounded by some constants. Hence, we conclude the following lemma:

Lemma 1 ([1,10]). *Let (S, δ) be a regular FGS and p a unary predicate symbol. Given a ground term graph g, the problem of deciding whether or not $g \in GL(S, \delta, p)$ is computed in $O((3^d n)^{w+1})$ time, where n is the number of vertices of g, d is the maximum degree of g, and w is the maximum treewidth of the term graphs in the heads of graph rewriting rules in S.*

3 Learning Regular FGS with 1-Finite Context Property

We consider $\mathcal{G}_{d,w}(\Sigma, \Lambda)$ as a universal set ($d, w \geq 0$). A positive presentation of a nonempty graph language $L \subseteq \mathcal{G}_{d,w}(\Sigma, \Lambda)$ is an infinite sequence $g_1, g_2, \ldots$ of elements in L such that $\{g_1, g_2, \ldots\} = L$.

An inductive inference machine (IIM, for short) is an effective procedure, or a certain type of Turing machine, which outputs a regular FGS and a predicate symbol each time a ground term graph is given. Let $L_* \subseteq \mathcal{G}_{d,w}(\Sigma, \Lambda)$ be a target graph language. We assume that an IIM has an access to an oracle Mem_{L_*} who answers membership queries. The query asks whether or not an arbitrary ground term graph g is included in L_*. Let $\tau = g_1, g_2, \ldots$ be a positive presentation of

L_*. An IIM outputs a regular FGS (S_i, δ_i) and a predicate symbol p_i by using membership queries each time a ground term graph g_i in τ is given. An IIM is said to *converge to a regular FGS (S, δ) and a predicate symbol p for τ with polynomial time update by using membership queries*, if M outputs a regular FGS (S_i, δ_i) and a predicate symbol p_i in polynomial time w.r.t. the sum of the size of the given ground term graphs so far, i.e., $|g_1| + |g_2| + \cdots + |g_i|$, and there exists a positive integer $n \geq 1$ with $(S_m, \delta_m) = (S, \delta)$ and $p_m = p$ for any $m \geq n$. Let $\mathcal{C} \subseteq 2^{\mathcal{G}_{d,w}(\Sigma, \Lambda)}$ be a class and $L_* \in \mathcal{C}$. A class $\mathcal{C}$ is said to be *identifiable in the limit with polynomial time update by using membership queries from positive data*, if there exists an IIM M such that for any $L_* \in \mathcal{C}$ and any presentation τ of L_*, M converges to a regular FGS (S, δ) and a predicate symbol p for τ with $GL(S, \delta, p) = L_*$ with polynomial time update by using membership queries.

Let $g = (V_g, E_g, \varphi_g, \psi_g, \emptyset, \emptyset, \emptyset)$ be a ground term graph and $\sigma = (v_1, \ldots, v_\ell)$ a list of distinct vertices in V_g $(1 \leq \ell \leq |V_g|)$. Let x be a new variable label in X that does not appear so far. For the ground term graph fragment $[g, \sigma]$, we denote by $g(\sigma)$ the term graph $(V_g, E_g, \varphi_g, \psi_g, \{h\}, \lambda_g, ports_g)$ where $h = \{v_1, \ldots, v_\ell\}$, $\lambda_g(h) = x$, and $ports_g(h) = \sigma$. In order to make the argument easier, we assume that g has no isolated vertex. Let $\{E_\alpha, E_\beta\}$ be a partition of E_g. Let V_α be the set of all endpoints of edges in E_α and V_β the set of all endpoints of edges in E_β. Let σ be one of the ordered lists of all vertices in $V_\alpha \cap V_\beta$. We obtain two ground term graph fragments $[\alpha, \sigma]$ and $[\beta, \sigma]$. We easily see that $\alpha(\sigma)\{x := [\beta, \sigma]\}$ and $\beta(\sigma)\{x := [\alpha, \sigma]\}$ are isomorphic to g.

For $[\alpha, \sigma_\alpha], [\beta, \sigma_\beta] \in \mathcal{F}(\Sigma, \Lambda)$, we define an operation $\odot$ as follows:

$$[\alpha, \sigma_\alpha] \odot [\beta, \sigma_\beta] = \begin{cases} \alpha(\sigma_\alpha)\{x := [\beta, \sigma_\beta]\} & \text{if } |\sigma_\alpha| = |\sigma_\beta|, \\ undefined & \text{otherwise.} \end{cases}$$

In Fig. 5, we give an example of α and β by a partition of E_g of a ground term graph g. Note that in general, $[\alpha, \sigma_\alpha] \odot [\beta, \sigma_\beta]$ is not always equivalent to $[\beta, \sigma_\beta] \odot [\alpha, \sigma_\alpha]$, because the vertex labels in the first operand always survive by any binding. If $\varphi_\alpha(\sigma_\alpha) = \varphi_\beta(\sigma_\beta)$, $[\alpha, \sigma_\alpha] \odot [\beta, \sigma_\beta] = [\beta, \sigma_\beta] \odot [\alpha, \sigma_\alpha]$ holds.

Let d and w be constant nonnegative integers. For a nonempty finite set of ground term graphs $D \subseteq \mathcal{G}_{d,w}(\Sigma, \Lambda)$, let

$$Sub(D) = \{[\beta, \sigma_\beta] \in \mathcal{F}_{d,w}(\Sigma, \Lambda) \mid \exists [\alpha, \sigma_\alpha] \in \mathcal{F}_{d,w}(\Sigma, \Lambda)[[\alpha, \sigma_\alpha] \odot [\beta, \sigma_\beta] \in D]\},$$
$$Con(D) = \{[\alpha, \sigma_\alpha] \in \mathcal{F}_{d,w}(\Sigma, \Lambda) \mid \exists [\beta, \sigma_\beta] \in \mathcal{F}_{d,w}(\Sigma, \Lambda)[[\alpha, \sigma_\alpha] \odot [\beta, \sigma_\beta] \in D]\}.$$

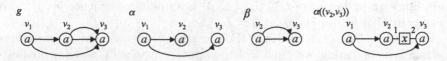

Fig. 5. Two ground term graphs α and β obtained from g by a partition of E_g: It is easy to see that $\alpha((v_2, v_3))\{x := [\beta, (v_2, v_3)]\}$ is isomorphic to g.

We give an example of $Con(D)$ in Fig. 6. It is easy to see that $Con(D) \subseteq Sub(D)$ holds. For any $[\beta, \sigma_\beta] \in Sub(D) \backslash Con(D)$, there is a ground term graph fragment $[\alpha, \sigma_\alpha] \in Con(D)$ such that α is isomorphic to β via an isomorphism ξ with $\xi(\sigma_\alpha) = \sigma_\beta$, ignoring the vertex labels in σ_α. Thus, unlike string grammars, we only use $Con(D)$ to learn a target regular FGS language from positive data. We have the following proposition:

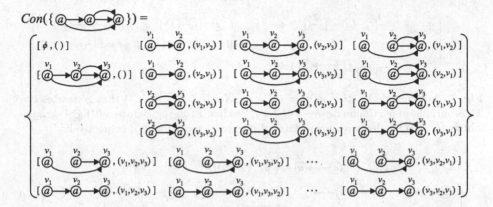

Fig. 6. An example of $Con(D)$.

Proposition 1. *Let D be a nonempty finite subset of $\mathcal{G}_{d,w}(\Sigma, \Lambda)$. Both $|Sub(D)|$ and $|Con(D)|$ are of polynomial size w.r.t. $\sum_{g \in D} |g|$.*

Let $(S, \delta) = ((\Sigma, \Lambda, X, \Pi, \Gamma), \delta)$ be a regular FGS. For a term graph f and $q \in \Pi$, if there are distinct $|\delta(q)|$ vertices $v_1, \ldots, v_{|\delta(q)|}$ in V_f such that $(\varphi_f(v_1), \ldots, \varphi_f(v_{|\delta(q)|})) = \delta(q)$ and for any $v \in V_f \backslash \{v_1, \ldots, v_{|\delta(q)|}\}$, $\varphi_f(v)$ is not a member of $\delta(q)$, we define $\varphi_f^{-1}(\delta(q)) = (v_1, \ldots, v_{|\delta(q)|})$, otherwise we define $\varphi_f^{-1}(\delta(q)) = ()$. Let p, q in Π and $[g, \sigma_g]$ in $\mathcal{F}_{d,w}(\Sigma, \Lambda)$. We define

$$C(S, \delta, p, q, [g, \sigma_g]) = \{f \in \mathcal{G}_{d,w}(\Sigma, \Lambda) \mid [g, \sigma_g] \odot [f, \varphi_f^{-1}(\delta(q))] \in GL(S, \delta, p)\}.$$

Definition 7 (1-FCP). *Let (S, δ) be a regular FGS and p, q unary predicate symbols of S. A term graph fragment $[g, \sigma_g]$ is said to be a context of q w.r.t. (S, δ, p) if $C(S, \delta, p, q, [g, \sigma_g]) = GL(S, \delta, q)$ holds. We say that (S, δ, p) has the 1-finite context property (1-FCP) if every predicate $q \in \Pi$ has a context of it.*

For the ground term graphs α, β in Fig. 5, $[\alpha, (v_2, v_3)]$ is a context of $q_{(2,2)}$ and $[\beta, (v_2, v_3)]$ is a context of $q_{(1,1)}$ w.r.t. $(S_3^{(3)}, \delta_3^{(3)}, p)$ in Fig. 7. We give an example $C(S_3^{(3)}, \delta_3^{(3)}, p, q_{(2,1)}, [g, \sigma_g])$ in Fig. 8 for some $[g, \sigma_g]$. We easily see that for any $d \geq 2$, the regular FGS in Fig. 7 has the 1-finite context property (1-FCP).

Definition 8 (1-FCP regular FGS language class). *$1\text{-}FCP\text{-}\mathcal{RFGSL}(d, w)$ denotes the set of all regular FGS languages $L \subseteq \mathcal{G}_{d,w}(\Sigma, \Lambda)$ that satisfies the following conditions:*

$$\Sigma_3 = \{a, s, t\}, \Lambda_3 = \{\epsilon\}, X_3 = \{x, x_1, \ldots\}, \Pi_3^{(d)} = \{p\} \cup \{q_{(i,j)} \mid 1 \le i, j \le d\},$$
$$\delta_3^{(d)}(p) = (), \delta_3^{(d)}(q_{(i,j)}) = (s, t) \ (1 \le i, j \le d).$$

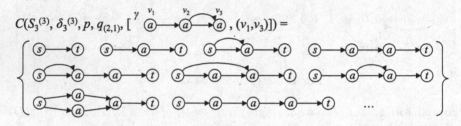

Fig. 7. A regular FGS $(S_3^{(d)}, \delta_3^{(d)}) = ((\Sigma_3, \Lambda_3, X_3, \Pi_3^{(d)}, \Gamma_3^{(d)}), \delta_3^{(d)})$ that generates the TTSP graphs of maximum degree d $(d \ge 2)$: Predicates $q_{(i,j)}$ generates all TTSP graphs whose vertices labeled with s and t are of degree at most i and j, respectively.

$$C(S_3^{(3)}, \delta_3^{(3)}, p, q_{(2,1)}, [\ \gamma \ @ \xrightarrow{v_1} @ \xrightarrow{v_2} @ \xrightarrow{v_3} , (v_1, v_3)]) =$$

Fig. 8. $C(S_3^{(3)}, \delta_3^{(3)}, p, q_{(2,1)}, [\gamma, (v_1, v_3)]) = GL(S_3^{(3)}, \delta_3^{(3)}, q_{(2,1)})$ holds, where $(S_3^{(3)}, \delta_3^{(3)})$ is a regular FGS in Fig. 7. Thus, $[\gamma, (v_1, v_3)]$ is a context of $q_{(2,1)}$ w.r.t. $(S_3^{(3)}, \delta_3^{(3)}, p)$.

1. L is definable by $(S, \delta, p) = ((\Sigma, \Lambda, X, \Pi, \Gamma), \delta, p)$ that has the 1-FCP,
2. Γ is written in Chomsky normal form, and
3. The treewidth of each term graph in Γ is at most w. Therefore, the maximum length of ports of the hyperedges in Γ is also at most $w + 1$.

Let $L_* \subseteq \mathcal{G}_{d,w}(\Sigma, \Lambda)$ be a target regular FGS language. We give a learning algorithm for 1-FCP-$\mathcal{RFGSL}(d, w)$ in Algorithm 1, which is a process of searching in $Con(D)$ for contexts of the predicate symbols in L_*. We construct a regular FGS $S(F, K) = (\Sigma, \Lambda, X, \Pi, \Gamma)$, pointer δ, and initial predicate p as follows:

- $\Sigma = \Sigma' \cup \{s_1, \ldots, s_{w+1}\}$, where $\Sigma' = \{a \mid \exists g_i \in D, \exists v \in V_{g_i}[\ \varphi_{g_i}(v) = a\ ]\}$ and $\Sigma' \cap \{s_1, \ldots, s_{w+1}\} = \emptyset$.
- $\Lambda = \{a \mid \exists g_i \in D, \exists e \in E_{g_i}[\ \psi_{g_i}(e) = a\ ]\}$.
- X: We use a new variable label only when needed.
- $\Pi = \{[\![\alpha, \sigma_\alpha]\!] \mid [\alpha, \sigma_\alpha] \in F \subseteq Con(D)\}$. Let $[\![\emptyset, ()]\!]$ be the initial predicate p.
- δ, Γ: In Table 1, we describe the pointer $\delta(q)$ for each predicate q in Π and the graph rewriting rules in Γ. In the table, we use the following notations.

Algorithm 1. Learn_1-FCP-$\mathcal{RFGSL}$

1: Let $K := \emptyset$, $F := \emptyset$;
2: **for** $n = 1, 2, 3, \ldots$ **do**
3:    Let $D = \{g_1, g_2, \ldots, g_n\}$;
4:    **if** $D \not\subseteq GL(S(F, K), \delta, p)$ **then** {By the parsing algorithm in [1].}
5:        Let $F := Con(D)$;
6:    **end if**
7:    Let $K := Con(D)$;
8:    output $(S(F, K), \delta, p)$;
9: **end for**

Let k, ℓ be two positive integers ($k \leq \ell$) and $\mathcal{P}_{k,\ell}$ the set of all list of k distinct positive integers that are less than or equal to ℓ. Let $\sigma = (\ell_1, \ldots, \ell_k) \in \mathcal{P}_{k,\ell}$. For a list of elements $\nu = (v_1, \ldots, v_\ell)$ ($k \leq \ell$), $\chi_\sigma(\nu)$ denotes $(v_{\ell_1}, \ldots, v_{\ell_k})$ and $\bar{\chi}_\sigma(\nu)$ denotes the list obtained from ν by deleting $v_{\ell_1}, \ldots, v_{\ell_k}$.

The graph rewriting rule R_1 in Fig. 9 is an example of the graph rewriting rules constructed by Table 1 for the target regular FGS language $GL(S_3^{(3)}, \delta_3^{(3)}, p)$.

Fig. 9. A graph rewriting rule R_1 constructed by the second table in Table 1: This graph rewriting rule corresponds to the rule R_2 of $(S_3^{(3)}, \delta_3^{(3)})$ in Fig. 7.

Theorem 1. *Let d and w be constant integers greater than zero. The class 1-FCP-$\mathcal{RFGSL}(d, w)$ is identifiable in the limit with polynomial time update by using membership queries from positive data.*

Proof. Let $(S_1, \delta_1, p_1), (S_2, \delta_2, p_2), \ldots, (S_i, \delta_i, p_i), \ldots$ be hypotheses output by Algorithm 1, and $(S_i, \delta_i, p_i) = ((\Sigma, \Lambda, X, \Pi_i, \Gamma_i), \delta_i, p_i)$. We prove that there exists a positive integer k such that $GL(S_n, \delta_n, p_n) = L_*$ for any integer $n \geq k$. Let $(S_*, \delta_*) = ((\Sigma, \Lambda, X, \Pi_*, \Gamma_*), \delta_*)$ be a regular FGS and p_* a predicate symbol in Π_* with $L_* = GL(S_*, \delta_*, p_*)$. Let G_i be a ground term graph given to Algorithm 1 at the i-th time, and $D_i = \{G_1, G_2, \ldots, G_i\}$. From the property of positive presentations, there exists a positive integer $n \geq 1$ such that $Con(D_n)$ has a ground term graph fragment $[g, \sigma_g]$ with $C(S_*, \delta_*, p_*, q, g) = GL(S_*, \delta_*, q)$ for any $q \in \Pi_*$. From the n-th input and after, for any predicate symbol $q \in \Pi_*$, Algorithm 1 has a ground term graph fragment corresponding to q. Thus, any graph rewriting rule in Γ_* is included in Γ_m for any $m \geq n$. It follows that $L_* \subseteq GL(S_m, \delta_m, p_m)$ for any $m \geq n$.

Table 1. $(S(F,K), \delta, p)$: There are three types of terminal rules and one type of binary rule. Each graph rewriting rule is created if the corresponding condition is satisfied. All conditions can be determined by asking to the membership oracle Mem_{L_*}. The unary rules can be constructed in a similar way to the binary rules. We omit its detail.

Terminal rules $p_0(f_0) \leftarrow$ in $(S(F,K), \delta, p)$:

p_0	$\delta(p_0)$	f_0	Condition
$[\![g_0, \sigma_{g_0}]\!]$ $\|\sigma_g\| = 1$	(s_1)	$(\{v_1\}, \emptyset, \varphi, \emptyset, \emptyset, \emptyset, ())$ $\varphi(v_1) = s_1$	$[g_0, \sigma_{g_0}] \odot [f_0, (v_1)] \in L_*$
$[\![g_0, \sigma_{g_0}]\!]$ $\|\sigma_g\| = 1$	(s_1)	$(\{v_1, v_2\}, \{(v_1, v_2)\}, \varphi, \psi, \emptyset, \emptyset, ())$ $\varphi(v_1) = s_1, \varphi(v_2) \in \Sigma'$	$[g_0, \sigma_{g_0}] \odot [f_0, (v_1)] \in L_*$
$[\![g_0, \sigma_{g_0}]\!]$ $\|\sigma_g\| = 2$	(s_1, s_2)	$(\{v_1, v_2\}, \{(v_1, v_2)\}, \varphi, \psi, \emptyset, \emptyset, ())$ $\varphi(v_1) = s_1, \varphi(v_2) = s_2$	$[g_0, \sigma_{g_0}] \odot [f_0, (v_1, v_2)] \in L_*$

Binary rules $p_1(f_1) \leftarrow p_2(f_2), p_3(f_3)$ in $(S(F,K), \delta, p)$

$p_i \ (i = 2,3)$	$\delta(p_i) \ (i = 2,3)$	$f_i \ (i = 2,3, j = 1, \ldots, \ell_i)$
$[\![g_i, \sigma_{g_i}]\!]$ $\|\sigma_{g_i}\| = \ell_i$	$(s_1, \ldots, s_{\ell_i})$	$f_i = (\{v_{i,1}, \ldots, v_{i,\ell_i}\}, \emptyset, \varphi_i, \emptyset, \{h_i\}, \lambda_i, port_i)$, where $\varphi_i(v_{i,j}) = s_j, \lambda_2(h_2) \neq \lambda_3(h_3), ports_i(h_i)[j] = v_{i,j}$.

p_1	$\delta(p_1)$	f_1
$[\![g_1, \sigma_{g_1}]\!]$ $\|\sigma_{g_1}\| = \ell_1$	$(s_1, \ldots, s_{\ell_1})$	$f_1 = [f_2, \chi_{\sigma_2}(ports_2(h_2))] \odot [f_3, \chi_{\sigma_3}(ports_3(h_3))]$, where $\sigma_i \in \mathcal{P}_{k, \|ports_i(h_i)\|} \ (i = 2,3)$ for some k. Let $\nu = ports_{f_2}(h_2) \cdot \bar{\chi}_{\sigma_3}(ports_{f_3}(h_3))$ and $\sigma_1 \in \mathcal{P}_{\ell_1, |\nu|}$. The vertices in ν are relabeled so that $\chi_{\sigma_1}(\nu) = (s_1, \ldots, s_{\ell_1})$ and $\bar{\chi}_\sigma(\nu) \in \Sigma'^{|\nu| - \ell_1}$.

Condition
For $\forall [\tau_2, \sigma_{\tau_2}], [\tau_3, \sigma_{\tau_3}] \in K$, if $[g_2, \sigma_{g_2}] \odot [\tau_2, \sigma_{\tau_2}] \in L_*$ and $[g_3, \sigma_{g_3}] \odot [\tau_3, \sigma_{\tau_3}] \in L_*$, then $[g_1, \sigma_{g_1}] \odot [[\tau_2, \chi_{\sigma_2}(\sigma_{\tau_2})] \odot [\tau_3, \chi_{\sigma_3}(\sigma_{\tau_3})], \xi] \in L_*$, where $\xi = \chi_{\sigma_1}(\chi_{\sigma_2}(\sigma_{\tau_2}) \cdot \bar{\chi}_{\sigma_3}(\sigma_{\tau_3}))$.

We assume that for any $n \geq 1$, there exists a positive integer $m \geq n$ such that $GL(S_m, \delta_m, p_m) \not\subseteq L_*$. Then there exists a ground term graph $G' \in GL(S_m, \delta_m, p_m) \backslash L_*$. Since $G' \in GL(S_m, \delta_m, p_m)$, there exist a graph rewriting rule $p_1(f_1) \leftarrow p_2(f_2), p_3(f_3)$ in Γ_m and ground term graph fragments $[\rho_2, \sigma_{\rho_2}]$ and $[\rho_3, \sigma_{\rho_3}]$ such that $[g_2, \sigma_{g_2}] \odot [\rho_2, \sigma_{\rho_2}] \in L_*$, $[g_3, \sigma_{g_3}] \odot [\rho_3, \sigma_{\rho_3}] \in L_*$ and G' is isomorphic to $[g_1, \sigma_{g_1}] \odot [[\rho_2, \chi_{\sigma_2}(\sigma_{\rho_2})] \odot [\rho_3, \chi_{\sigma_3}(\sigma_{\rho_3})], \xi]$, where p_1, p_2 and p_3 correspond to $[g_1, \sigma_{g_1}], [g_2, \sigma_{g_2}]$ and $[g_3, \sigma_{g_3}]$, respectively. There exist ground term graph fragments $[\tau_2, \sigma_{\tau_2}], [\tau_3, \sigma_{\tau_3}] \in K = Con(D_\ell)$ with $[g_2, \sigma_{g_2}] \odot [\tau_2, \sigma_{\tau_2}] \in L_*$, $[g_3, \sigma_{g_3}] \odot [\tau_3, \sigma_{\tau_3}] \in L_*$ and $[g_1, \sigma_{g_1}] \odot [[\tau_2, \chi_{\sigma_2}(\sigma_{\tau_2})] \odot [\tau_3, \chi_{\sigma_3}(\sigma_{\tau_3})], \xi] \notin L_*$ for some positive integer ℓ. Thus, $p_1(f_1) \leftarrow p_2(f_2), p_3(f_3)$ is removed from Γ_ℓ. This contradicts that $p_1(f_1) \leftarrow p_2(f_2), p_3(f_3)$ in Γ_m. Therefore, we can show that there exists a positive integer k such that $GL(S_n, \delta_n, p_n) = L_*$ for any integer $n \geq k$. From Proposition 1, $|Con(D_n)|$ is of polynomial size w.r.t. $\sum_{i=1}^n |G_i|$ at the n-th step. Thus, from Lemma 1, the n-th hypothesis (S_n, δ_n, p_n) is output by Algorithm 1 with polynomial update time w.r.t $\sum_{i=1}^n |G_i|$.   □

4 Conclusions

We have considered the problem of learning FGS languages from the viewpoint of the computational learning theory. First, we introduced the class 1-FCP-$\mathcal{RFGSL}$ of regular FGS languages of bounded degree and treewidth with 1-finite context property (1-FCP). We also presented an algorithm for learning class 1-FCP-$\mathcal{RFGSL}$ by using current distributional learning techniques [11]. Finally, we showed that class 1-FCP-$\mathcal{RFGSL}$ can be identifiable in the limit with polynomial time update by using membership queries from positive data. This result will lead us to develop new techniques for learning other classes of FGS languages with distributional properties.

Clark et al. [2,3] and Yoshinaka [11,12] discussed the learnabilities of the class of languages of context-free grammars with the finite kernel property (FKP) and finite context property (FCP). As future work, we will consider the polynomial time learnabilities of the class of regular FGS languages with the FKP and FCP.

References

1. Chiang, D., Andreas, J., Bauer, D., Hermann, K.M., Jones, B., Knight, K.: Parsing graphs with hyperedge replacement grammars. In: Proceeding of ACL 2013, pp. 924–932. Association for Computational Linguistic (2013)
2. Clark, A.: A learnable representation for syntax using residuated lattices. In: Groote, P., Egg, M., Kallmeyer, L. (eds.) FG 2009. LNCS, vol. 5591, pp. 183–198. Springer, Heidelberg (2011). doi:10.1007/978-3-642-20169-1_12
3. Clark, A., Eyraud, R.: Polynomial identification in the limit of substitutable context-free languages. J. Mach. Learn. Res. **8**, 1725–1745 (2007)
4. Drewes, F., Kreowski, H.J., Habel, A.: Hyperedge replacement graph grammars. In: Handbook of Graph Grammars and Computing by Graph Transformation, vol. 1, pp. 95–162. World Scientific (1997)
5. Gildea, D.: Grammar factorization by tree decomposition. Comput. Linguist. **37**(10), 231–248 (2011)
6. Hara, S., Shoudai, T.: Polynomial time MAT learning of c-deterministic regular formal graph systems. In: Proceeding IIAI-AAI 2014, pp. 204–211. IEEE (2014)
7. Kasprzik, A., Yoshinaka, R.: Distributional learning of simple context-free tree grammars. In: Kivinen, J., Szepesvári, C., Ukkonen, E., Zeugmann, T. (eds.) ALT 2011. LNCS (LNAI), vol. 6925, pp. 398–412. Springer, Heidelberg (2011). doi:10.1007/978-3-642-24412-4_31
8. Lautemann, C.: The complexity of graph languages generated by hyperedge replacement. Acta Informatica **27**(5), 399–421 (1990)
9. Okada, R., Matsumoto, S., Uchida, T., Suzuki, Y., Shoudai, T.: Exact learning of finite unions of graph patterns from queries. In: Hutter, M., Servedio, R.A., Takimoto, E. (eds.) ALT 2007. LNCS (LNAI), vol. 4754, pp. 298–312. Springer, Heidelberg (2007). doi:10.1007/978-3-540-75225-7_25
10. Uchida, T., Shoudai, T., Miyano, S.: Parallel algorithms for refutation tree problem on formal graph systems. IEICE Trans. Inf. Syst. **E78–D**(2), 99–112 (1995)

11. Yoshinaka, R.: Integration of the dual approaches in the distributional learning of context-free grammars. In: Dediu, A.-H., Martín-Vide, C. (eds.) LATA 2012. LNCS, vol. 7183, pp. 538–550. Springer, Heidelberg (2012). doi:10.1007/978-3-642-28332-1_46
12. Yoshinaka, R.: General perspective on distributionally learnable classes. In: Proceeding of MoL 2015, pp. 87–98. Association for Computational Linguistic (2015)

Learning Relational Dependency Networks for Relation Extraction

Ameet Soni[1]([⊠]), Dileep Viswanathan[2], Jude Shavlik[3],
and Sriraam Natarajan[2]

[1] Swarthmore College, Swarthmore, PA, USA
soni@cs.swarthmore.edu
[2] Indiana University, Bloomington, IN, USA
{diviswan,natarasr}@indiana.edu
[3] University of Wisconsin, Madison, WI, USA
shavlik@cs.wisc.edu

Abstract. We consider the task of KBP slot filling – extracting relation information from newswire documents for knowledge base construction. We present our pipeline, which employs Relational Dependency Networks (RDNs) to learn linguistic patterns for relation extraction. Additionally, we demonstrate how several components such as weak supervision, word2vec features, joint learning and the use of human advice, can be incorporated in this relational framework. We evaluate the different components in the benchmark KBP 2015 task and show that RDNs effectively model a diverse set of features and perform competitively with current state-of-the-art relation extraction methods.

1 Introduction

The problem of knowledge base population (KBP) – constructing a knowledge base (KB) of facts gleaned from a large corpus of unstructured data – poses several challenges for the NLP community. Commonly, this relation extraction task is decomposed into two subtasks – entity linking, in which entities are linked to already identified identities within the document or to entities in the existing KB, and slot filling, which identifies certain attributes about a target entity.

We present our system for KBP slot filling based on probabilistic logic formalisms and present the different components of the system. Specifically, we employ Relational Dependency Networks [14], a formalism that has been successfully used for joint learning and inference from stochastic, noisy, relational data. We consider our RDN system against the current state-of-the-art for KBP to demonstrate the effectiveness of our probabilistic relational framework. Additionally, we show how RDNs can effectively incorporate many popular approaches in relation extraction such as joint learning, weak supervision, word2vec features, and human advice, among others.

We provide a comprehensive comparison of various settings such as joint learning vs learning of individual relations, use of weak supervision vs gold standard labels, using expert advice vs only learning from data, etc. These questions

© Springer International Publishing AG 2017
J. Cussens and A. Russo (Eds.): ILP 2016, LNAI 10326, pp. 81–93, 2017.
DOI: 10.1007/978-3-319-63342-8_7

are extremely interesting from a general machine learning perspective, but also critical to the NLP community. As we show empirically, the key contributions of this paper are as follows:

- Our RDN framework is competitive, and often superior, to state-of-the-art systems for KBP slot filling.
- RDNs successfully incorporate various types of features, including advice, joint learning, and word2vec features.
- Ours is the first KBP system to leverage *knowledge-based weak supervision* – a logic-based framework that we have previously shown to be complementary and often superior to distant supervision.

Some of the results such as human advice being useful in many relations and joint learning being beneficial in the cases where the relations are correlated among themselves are on the expected lines. However, some surprising observations include the fact that weak supervision and word2vec features are not as useful as expected, although further investigation is warranted.

We first present the proposed pipeline with all the different components of the learning system. Next we present the set of 14 relations that we learn on before presenting the experimental results. We finally discuss the results of these comparisons before concluding by presenting directions for future research.

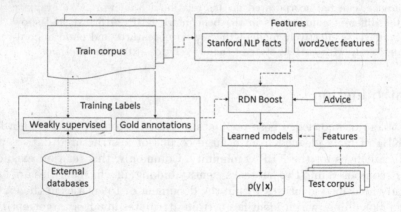

Fig. 1. Pipeline Full RDN relation extraction pipeline

2 Background

As a part of the Text Analysis Conference (TAC), NIST has supported several tasks related to Knowledge Base Population (KBP) including English Slot Filling [23]. The goal of this task is to mine a corpus of text data (e.g., newswire articles) for information on two specific categories of entities – persons and organizations. The type of information is predefined as relations (e.g., $parent(a, b)$ specifies that person b is a parent of person a).

Over the last several years, many approaches have been proposed across a spectrum of machine learning approaches. A common thread to these approaches

Table 1. Standard NLP Features: Features derived from the training corpus used by our learning system. POS - part of speech. NE - Named Entity. DPR - root of dependency path tree.

Feature	Description
wordString	Word with word id
wordPosition	Location of the word
caselessWordString	Word string in lower case
wordLemma	Canonical form of word
isNEWord	Whether word is NE
nextWords	Two succeeding words
prevWords	Two preceding words
nextPOS	POS for the succeeding words
prevPOS	POS for the preceding words
nextLemmas	Canonical form of successors
prevLemmas	Canonical form of predecessors
nextNE	Succeeding NE phrases
prevNE	Preceding NE phrases
lemmaBetween	Canonical form of word occurring between two NEs
neBetween	Word b/w two NEs is an NE
posBetween	POS of word b/w two NEs
Dependency Path	
rootChildLemma	Canonical form of child of DPR
rootChildNER	Child of DPR is NE
rootChildPOS	POS of child of DPR
rootLemma	Lemma of DPR
rootNER	DPR is NER
rootPOS	POS of DPR

is the use of distant supervision [11]. The 2014 winner, DeepDive [17], leveraged Markov Logic Networks [2] with distant supervision to perform slot filling. RelationFactory [20] also utilizes distant supervision to train a highly modular pipeline that focuses on scaling and efficiency, employing several shallow classifiers (e.g., manually created patterns, learned rules, SVMs) for various tasks. Other approaches include multiple instance learning [10] and stacked ensembles which combine multiple submissions into a single framework [25].

3   Proposed Pipeline

We present the different aspects of our pipeline, depicted in Fig. 1.

3.1   Feature Generation

Given a training corpus of raw text documents, our learning algorithm first converts these documents into a set of facts (i.e., features) that are encoded in first order logic (FOL). Raw text is processed using the Stanford CoreNLP

Table 2. Rules for KB Weak Supervision: A sample of knowledge-based rules for weak supervision. The first value defines a weight, or confidence in the accuracy of the rule. The target relation appears at the end of each clause. "PER", "ORG", "NUM" represent entities that are persons, organizations, and numbers, respectively.

Weight	MLN Clause
1.0	entityType(a, "PER"), entityType(b, "NUM"), nextWord(a, c), word(c, ","), nextWord(c, b) → age(a, b)
0.6	entityType(a, "PER"), entityType(b, "NUM"), prevLemma(b, "age") → age(a, b)
0.8	entityType(a, "PER"), entityType(b, "PER"), nextLemma(a, "mother") → parents(a, b)
0.8	entityType(a, "PER"), entityType(b, "PER"), nextLemma(a, "father") → parents(a, b)

Toolkit[1] [6] to extract parts-of-speech, word lemmas, etc. as well as generate parse trees, dependency graphs and named-entity recognition information. The full set of extracted features is listed in Table 1. These are then converted into features in prolog format and are given as input to the system.

In addition to the structured features from the output of Stanford toolkit, we also use deeper features based on `word2vec` [8] as input to our learning system. Standard NLP features tend to treat words as individual objects, ignoring links between words that occur with similar meanings or, importantly, similar contexts (e.g., city-country pairs such as *Paris – France* and *Rome – Italy* occur in similar contexts). `word2vec` provide a continuous-space vector embedding of words that, in practice, capture many of these relationships [8,9]. We use word vectors from Stanford[2] and Google[3].

We generated features from word vectors by finding words with high similarity in the embedded space. That is, we used word vectors by considering relations of the following form: $isCosSimilar(wordA, relationB, threshold)$, if a word has a high cosine similarity to any keyword (e.g., "father") for a particular relation (e.g., *parent*). Details can be found in Sect. 4.3. At a high level, these types of features would allow our learner to generate rules that connect unique words that occur in similar contexts (e.g., "husband" and "wife"). Standard features, instead, would require the same rule to be learned multiple times (e.g., once each for "husband", "wife", "partner", etc. as in Fig. 2).

3.2   Weak Supervision

One difficulty with the KBP task is that very few documents come labeled with *gold standard labels*, and human annotation is prohibitively expensive beyond

---

[1] http://stanfordnlp.github.io/CoreNLP/.

[2] http://nlp.stanford.edu/projects/glove/.

[3] https://code.google.com/p/word2vec/.

a few hundred documents. This is problematic for discriminative learning algorithms which excel when given a large supervised training corpus. To overcome this obstacle, we employ *weak supervision* – the use of external knowledge (e.g., a database) to heuristically label examples. Following our work in Soni et al. [21], we employ our novel knowledge-based weak supervision approach, as opposed to the more traditional distant supervision which references an external database of known relations.

Knowledge-based weak supervision is based on previous work [13,21] with the following insight: labels are typically created by "domain experts" who annotate the labels carefully, and who typically employ some inherent rules in their mind to create examples. For example, when identifying family relationship, we may have an *inductive bias* towards believing two persons in a sentence with the same last name are related, or that the words "son" or "daughter" are strong indicators of a parent relation. We call this *world knowledge* as it describes the domain (or the world) of the target relation.

For the KBP task, some rules that we used are shown in Table 2. For example, the first rule identifies any number following a person's name and separated by a comma is likely to be the person's age (e.g., "Sharon, 42"). The third and fourth rule provide examples of rules that utilize more textual features; these rules state the appearance of the lemma "mother" or "father" between two persons is indicative of a parent relationship. Previous results show this approach produces more examples with less overhead than distant supervision and can be employed where relevant database are not available.

To this effect, we encode the domain expert's knowledge in the form of first-order logic rules with accompanying weights to indicate the expert's confidence. We use the probabilistic logic formalism *Markov Logic Networks* [2] to perform inference on unlabeled text (e.g., the TAC KBP corpus). Potential entity pairs from the corpus are queried to the MLN, yielding (weakly-supervised) positive examples. We choose MLNs as they permit domain experts to easily write rules while providing a probabilistic framework that can handle noise, uncertainty, and preferences. We use the Tuffy system [16] to perform inference as it is robust and scales well to millions of documents[4].

3.3 Learning Relational Dependency Networks

Previous research [7] has demonstrated that joint inferences of the relations are more effective than considering each relation individually. Consequently, we have considered a formalism that has been successfully used for joint learning and inference from stochastic, noisy, relational data called Relational Dependency Networks (RDNs) [12,14]. RDNs extend dependency networks (DN) [4] to the relational setting. The key idea in a DN is to approximate the joint distribution over a set of random variables as a product of their marginal distributions, i.e., $P(y_1, ..., y_n | \mathbf{X}) \approx \prod_i P(y_i | \mathbf{X})$. It has been shown that employing Gibbs

---

[4] As the structure and weights are predefined by the expert, learning is not needed for our MLN.

Table 3. Advice Rules: Sample advice rules used for relation extraction. We employed a total of 72 such rules for our 14 relations.

Advice Rules
Entity preceded by a number and a phrase "year-old" probably refers to age.
Entity present with a phrase in sentence "who turned" probably refers to age.
Entity1 is "also known as" Entity2 probably refers to alternate name.
Entity1, "nicknamed" Entity2 probably refers to alternate name.
Entity1 followed by phrase "is a citizen of" Entity2 probably refers to origin.
Entity followed by phrase "is a devout" Entity2 probably refers to religion.
Entity, followed by "a" Entity2"-based company" probably refers to city/state/country of headquarters.
If Entity1 and Entity2 are siblings then they are not parents of each other.
If Entity1 and Entity2 are spouses of each other then they are not parents of each other

sampling in the presence of a large amount of data allows this approximation to be particularly effective. Note that, one does not have to explicitly check for acyclicity making these DNs particularly easy to be learned. We refer the reader to previous work [12,14] for more details and examples of the RDN model.

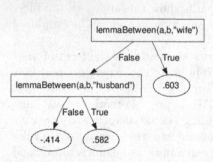

Fig. 2. Example regression tree for the *siblings* relation. This tree states that the weight for the relation being true is higher if either "husband" or "wife" appear between the entities.

In an RDN, typically, each distribution is represented by a relational probability tree (RPT) [15]. However, following previous work [12], we replace the RPT of each distribution with a set of relational regression trees [1] built in a sequential manner i.e., replace a single tree with a set of gradient boosted trees. This approach, RDN Boost, has been shown to have state-of-the-art results in learning RDNs. An simplified regression tree for the siblings relation is provided in Fig. 2. Several boosted trees are learned for each relation and combined in ensemble fashion during inference.

3.4 Incorporating Human Advice

While most relational learning methods restrict the human to merely annotating the data, we go beyond and request the human for advice. The intuition is that we as humans read certain patterns and use them to deduce the nature of the relation between two entities present in the text. The goal of our work is to capture such mental patterns of the humans as advice to the learning algorithm. We modified the work of Odom et al. [18,19] to learn RDNs in the presence of advice. The key idea is to explicitly represent advice in calculating gradients. This allows the system to trade-off between data and advice throughout the learning phase, rather than only consider advice in initial iterations. Advice, in particular, become influential in the presence of noisy or less amount of data.

A few sample advice rules in English (converted to FOL in RDN Boost) are presented in Table 3. Note that some of the rules are "soft" rules in that they are not true in many situations. Odom et al. [19] weigh the effect of the rules against the data and hence allow for partially correct rules.

4   Experiments and Results

Table 4. Relations: The relations considered from TAC KBP. Columns indicate the number of training examples utilized – both human annotated (Gold) and weakly supervised (WS), when available – from TAC KBP 2014 and number of test examples from TAC KBP 2015. 10 relations describe person entities (*per*) while the last 4 describe organizations (*org*).

Relation	Gold	WS	Test
per : age	89	750	44
per : alternateName	28	x	18
per : children	89	x	23
per : origin	96	750	48
per : otherFamily	72	750	10
per : parents	71	750	30
per : religion	70	750	11
per : siblings	77	750	31
per : spouse	66	750	28
per : title	158	x	39
org : cityHQ	69	x	10
org : countryHQ	69	21	29
org : dateFounded	70	750	17
org : foundedBy	62	750	32

We now present our experimental evaluation. We considered 14 specific relations from two categories, *person* and *organization* from the TAC KBP competition. The relations considered are listed in the left column of Table 4. We utilize documents from KBP 2014 for training while utilizing a non-overlapping set of documents from the 2015 corpus for testing.

All RDN results presented are obtained from 5 different runs of the train and test sets to provide more robust estimates of accuracy[5]. We consider three standard metrics – area under the ROC curve, F-1 score and the recall at a certain precision. The train/test gold-standard sizes are provided in the table, including weakly supervised examples. Negative examples are created by randomly selecting paired entities in the same sentence (per relation and per run). We chose the precision as 0.66 since the fraction of positive examples to negatives is 1:2.

To analyze our system, we aimed to answer the following questions:

Q1: Do weakly supervised examples help construct better models?
Q2: Does joint learning help in some relations?
Q3: Are `word2vec` features more predictive than standard features?

---

[5] Please see [12] for standard settings. 25 trees were learned per relation per run, with maximum depth of 3 and advice learning rate of 0.25.

Q4: Does advice improve performance compared to just learning from data?
Q5: Does our system perform competitively against a robust baseline?

4.1   Weak Supervision

Table 5. Weak Supervision: Results comparing models trained with gold standard examples only (G) and models trained with gold standard and weakly supervised examples combined (G+WS).

Relation	AUC ROC		
	G	W	G+W
age	0.94	0.83	0.91
origin	0.88	0.69	0.77
otherFamily	0.88	0.88	0.93
parents	0.74	0.69	0.70
religion	0.77	0.70	0.80
siblings	0.82	0.72	0.74
spouse	0.86	0.86	0.76
countryHQ	0.79	0.60	0.79
dateFounded	0.87	0.81	0.84
foundedBy	0.85	0.61	0.70

To address **Q1**, we sought to analyze whether weakly supervised examples could substitute for a large gold-standard training set. Specifically, we evaluated 10 relations as show in Table 5. Based on experiments from [21], we utilized our knowledge-based weak supervision approach to provide positive examples, with a range of 4 to 8 rules for each relation. We compared three conditions: using (1) a large gold-standard training set (G), (2) a large weakly supervised data set (750 positive examples per relation) (W), and (3) using a small sample of 30 gold standard combined with 150 weakly supervised examples (G+W).

The results are presented in Table 5. With a few exceptions, a small set of seed gold standard examples combined with weakly supervised data mimic the results of a much larger (and time-consuming) gold-standard set of training data. The exceptions are cases where our knowledge-base approach struggles to find quality examples. In previous work, we showed that distant supervision may be better in cases like this where general world knowledge is difficult to encapsulate in rules [21]. Surprisingly, a large weak supervision set by itself does not seem to help learn better models for inferring relations in most cases. We hypothesize that the number of gold standard examples provided may be sufficient to learn RDN models. Thus **Q1** is answered equivocally, with weak supervision being able to supplement a small amount of gold examples.

4.2   Joint Learning

To address our next question, we assessed our pipeline when learning relations independently (i.e., individually) versus learning relations jointly within the RDN, displayed in Table 6. Recall and F1 are omitted for conciseness – the conclusions are the same across all metrics. Joint learning appears to help in about half of the relations (8/14). Particularly, in person category, joint learning with gold standard outperforms their individual learning counterparts. This is due to the fact that some relations such as parents, spouse, siblings etc. are inter-related and learning them jointly indeed improves performance. Hence **Q2** can be answered affirmatively for connected relations.

4.3 Word2vec

Table 7 shows the results of experiments comparing the RDN framework with and without `word2vec` features. We set the modes [22] such that the first argument to *isCosSimilar* is a '-' (i.e., existing) variable. We provide lists of candidate constants (on average a few dozen for each target relation) for the second argument using our knowledge of each target concept. For example, we include the words "father" and "mother" (for the *parent* relation) or "devout","convert", and "follow" (*religion* relation). For the third argument, we utilize a threshold for cosine similarity of 0.70 (i.e., a word is considered to be similar to a keyword if their cosine similarity is above 0.70).

   `word2vec` appears to largely have no impact on results. One possibility may be that this is due to a limitation in the depth of trees learned. Learning more and/or deeper trees may improve use of `word2vec` features. Moreover, our experiments were limited to a single threshold for similarity. Instead, we could provide a list of thresholds so that the learner could utilize different similarity thresholds for different contexts. **Q3** is answered cautiously in the negative, although future work could lead to improvements.

Table 6. Joint Learning: Results comparing relation models learned individually (IL) and jointly (JL).

Relation	AUC ROC	
	IL	JL
age	0.93	0.93
alternateName	0.91	0.75
children	0.75	0.76
origin	0.86	0.89
otherFamily	0.88	0.89
parents	0.74	0.74
religion	0.72	0.79
siblings	0.79	0.80
spouse	0.86	0.87
title	0.90	0.89
cityHQ	0.74	0.73
countryHQ	0.75	0.79
dateFounded	0.87	0.86
foundedBy	0.83	0.86

Table 7. word2vec: Results comparing models trained without (−w2v) and with `word2vec` features (+w2v).

Relation	AUC ROC	
	−w2v	+w2v
age	0.94	0.94
alternateName	0.75	0.73
children	0.76	0.79
origin	0.88	0.86
otherFamily	0.88	0.88
parents	0.74	0.76
religion	0.77	0.79
siblings	0.82	0.79
spouse	0.86	0.82
title	0.89	0.90
cityHQ	0.73	0.73
countryHQ	0.79	0.78
dateFounded	0.87	0.88
foundedBy	0.85	0.84

4.4   Advice

Table 8 shows the results of experiments that test the use of advice within the joint learning setting. The use of advice improves or matches the performance of using only joint learning. The key impact of advice can be mostly seen in the improvement of recall in several relations. It appears that in cases where there are not good quality examples, advice improves recall but in cases where there are already reasonable examples, advice does not improve the performance significantly. This is line with previous findings in other domains [3,5,19,24]. As this claim warrants further investigation, Q4 can be answered optimistically.

4.5   RDN Boost vs RelationFactory

Table 8. Advice: Results comparing models trained without (-Adv) and with advice (+Adv).

Relation	AUC ROC		Recall	
	−Adv	+Adv	−Adv	+Adv
age	0.93	0.93	0.56	0.74
alternateName	0.75	0.77	0.20	0.16
children	0.76	0.76	0.04	0.14
origin	0.89	0.88	0.86	0.82
otherFamily	0.89	0.90	0	0.06
parents	0.74	0.72	0.15	0.05
religion	0.79	0.81	0.51	0.56
siblings	0.80	0.81	0.04	0.00
spouse	0.87	0.85	0.06	0.04
title	0.89	0.90	0.16	0.07
cityHQ	0.73	0.74	0.26	0.28
countryHQ	0.79	0.77	0.61	0.62
dateFounded	0.86	0.86	0.20	0.05
foundedBy	0.86	0.84	0.24	0.25

RelationFactory (RF) [20] is an open-source system for performing relation extraction based on distantly supervised classifiers. It was the top system in the TAC KBP 2013 competition [23] and thus serves as a suitable baseline for our method. RF is very conservative in its responses, making it difficult to adjust the precision levels. To be most generous to RF, we present recall for all returned results. The AUC ROC, recall, and F1 scores of our system against RF are presented in Table 9. Inference in RF took approximately 15 min on a single CPU for the entire test set and 3 min for RDNs. Training and RDN took approximately 30 min per relation per run (RF is available pre-trained).

Based on the results, we can conclude for **Q5** that RDNs performs comparably, and often better than the state-of-the-art RelationFactory system. In particular, our method outperforms RelationFactory in AUC ROC across all relations. Recall is a mixed picture with both approaches showing some improvements – RDN outperforms in 6 relations while RelationFactory does so in 8. Note that in the instances where RDN provides superior recall, it does so with dramatic improvements (RF often returns 0 positives in these relations). F1 also shows RDN's superior performance, outperforming RF in most relations.

Table 9. RelationFactory (RF) vs RDN: Values in bold indicate superiour performance against the alternative approach.

Relation	AUC ROC		Recall		F1	
	RF	RDN	RF	RDN	RF	RDN
age	0.64	**0.93**	0.28	**0.74**	0.44	**0.67**
alternateName	0.50	**0.77**	0.00	**0.16**	0	**0.10**
children	0.54	**0.76**	0.09	**0.14**	0.17	**0.28**
origin	0.50	**0.89**	0.00	**0.86**	0	**0.64**
otherFamily	0.56	**0.90**	**0.11**	0.06	**0.24**	0.22
parents	0.29	**0.74**	**0.33**	0.15	**0.50**	0.31
religion	0.50	**0.81**	0	**0.56**	0	**0.60**
siblings	0.13	**0.81**	**0.17**	0.00	0.29	0.29
spouse	0.57	**0.85**	**0.13**	0.04	0.23	**0.37**
title	0.67	**0.90**	**0.67**	0.07	**0.80**	0.54
cityHQ	0.38	**0.74**	**0.38**	0.28	**0.55**	0.41
countryHQ	0.57	**0.77**	0.14	**0.62**	0.25	**0.58**
dateFounded	0.67	**0.86**	**0.33**	0.05	**0.50**	0.46
foundedBy	0.20	**0.84**	**0.37**	0.25	0.54	**0.55**

5   Conclusion

We presented our fully relational system utilizing Relational Dependency Networks for the Knowledge Base Population task. We demonstrated RDN's ability to effectively learn the relation extraction task, performing comparably (and often better) than the state-of-art RelationFactory system. Furthermore, we demonstrated the ability of RDNs to incorporate various concepts in a relational framework, including `word2vec`, human advice, joint learning, and weak supervision. While weak supervision did not significantly improve performance on its own, we demonstrate that our knowledge-base weak supervision approach is a viable alternative to the more popular distant supervision approach and can effectively substitute for a much larger (and expensive) gold-standard data set. Furthermore, while initial results show `word2vec` features do not improve accuracy on this task, our work does demonstrate a viable formulation for utilizing these features in a relation setting. Future directions include considering a larger number of relations, deeper features and finally, comparisons with more systems which have experienced recent success, such as DeepDive [17].

Acknowledgements. Dileep Viswanathan, Jude Shavlik and Sriraam Natarajan gratefully acknowledge the support of the DARPA DEFT Program under the Air Force Research Laboratory (AFRL) prime contract no. FA8750-13-2-0039. Any opinions, findings, and conclusion or recommendations expressed in this material are those of the authors and do not necessarily reflect the view of the DARPA, ARO, AFRL, or the US government.

References

1. Blockeel, H., Raedt, L.D.: Top-down induction of first-order logical decision trees. Artif. Intell. **101**(1), 285–297 (1998)
2. Domingos, P., Lowd, D.: Markov Logic: An Interface Layer for AI. Morgan & Claypool, San Rafael (2009)
3. Fung, G.M., Mangasarian, O.L., Shavlik, J.W.: Knowledge-based support vector machine classifiers. In: NIPS, pp. 01–09. MIT Press (2002)
4. Heckerman, D., Chickering, D., Meek, C., Rounthwaite, R., Kadie, C.: Dependency networks for inference, collaborative filtering, and data visualization. J. Mach. Learn. Res. **1**, 49–75 (2001)
5. Kunapuli, G., Odom, P., Shavlik, J., Natarajan, S.: Guiding an autonomous agent to better behaviors through human advice. In: ICDM, pp. 409–418 (2013)
6. Manning, C., Surdeanu, M., Bauer, J., Finkel, J., Bethard, S., McClosky, D.: The Stanford CoreNLP natural language processing toolkit. In: ACL, pp. 55–60 (2014)
7. Meza-Ruiz, I., Riedel, S.: Jointly identifying predicates, arguments and senses using Markov Logic. In: NAACL-HLT, pp. 155–163 (2009)
8. Mikolov, T., Chen, K., Corrado, G., Dean, J.: Efficient estimation of word representations in vector space. In: Workshop at ICLR (2013)
9. Mikolov, T., Yih, W., Zweig, G.: Linguistic regularities in continuous space word representations. In: NAACL-HLT, pp. 746–751 (2013)
10. Min, B., Grishman, R., Wan, L., Wang, C., Gondek, D.: Distant supervision for relation extraction with an incomplete knowledge base. In: NAACL, pp. 777–782 (2013)
11. Mintz, M., Bills, S., Snow, R., Jurafsky, D.: Distant supervision for relation extraction without labeled data. In: ACL, pp. 1003–1011 (2009)
12. Natarajan, S., Khot, T., Kersting, K., Gutmann, B., Shavlik, J.: Boosting relational dependency networks. In: ILP (2010)
13. Natarajan, S., Picado, J., Khot, T., Kersting, K., Re, C., Shavlik, J.: Effectively creating weakly labeled training examples via approximate domain knowledge. In: Davis, J., Ramon, J. (eds.) ILP 2014. LNCS (LNAI), vol. 9046, pp. 92–107. Springer, Cham (2015). doi:10.1007/978-3-319-23708-4_7
14. Neville, J., Jensen, D.: Relational dependency networks. In: Introduction to Statistical Relational Learning. The MIT Press (2007)
15. Neville, J., Jensen, D., Friedland, L., Hay, M.: Learning relational probability trees. In: SIGKDD, pp. 625–630 (2003)
16. Niu, F., Ré, C., Doan, A., Shavlik, J.W.: Tuffy: scaling up statistical inference in Markov logic networks using an RDBMS. VLDB **4**(6), 373–384 (2011)
17. Niu, F., Zhang, C., Ré, C., Shavlik, J.W.: DeepDive: web-scale knowledge-base construction using statistical learning and inference. VLDS **12**, 25–28 (2012)

18. Odom, P., Bangera, V., Khot, T., Page, D., Natarajan, S.: Extracting adverse drug events from text using human advice. In: Holmes, J.H., Bellazzi, R., Sacchi, L., Peek, N. (eds.) AIME 2015. LNCS, vol. 9105, pp. 195–204. Springer, Cham (2015). doi:10.1007/978-3-319-19551-3_26
19. Odom, P., Khot, T., Porter, R., Natarajan, S.: Knowledge-based probabilistic logic learning. In: AAAI, pp. 3564–3570 (2015)
20. Roth, B., Barth, T., Chrupala, G., Gropp, M., Klakow, D.: RelationFactory: a fast, modular and effective system for knowledge base population. In: EACL, pp. 89–92 (2014)
21. Soni, A., Viswanathan, D., Pachaiyappan, N., Natarajan, S.: A comparison of weak supervision methods for knowledge base construction. In: Automated Knowledge Base Construction (AKBC) Workshop at NAACL (2016)
22. Srinivasan, A.: The aleph manual (2001). http://www.comlab.ox.ac.uk/oucl/~research/areas/machlearn/Aleph/
23. Surdeanu, M.: Overview of the TAC 2013 knowledge base population evaluation: English slot filling and temporal slot filling. In: Text Analysis Conference (2013)
24. Towell, G., Shavlik, J.: Knowledge-based artificial neural networks. Artif. Intell. **70**(1–2), 119–165 (1994)
25. Viswanathan, V., Rajani, N.F., Bentor, Y., Mooney, R.J.: Stacked ensembles of information extractors for knowledge-base population. In: ACL, pp. 177–187 (2015)

Towards Nonmonotonic Relational Learning from Knowledge Graphs

Hai Dang Tran[1], Daria Stepanova[1(✉)], Mohamed H. Gad-Elrab[1],
Francesca A. Lisi[2], and Gerhard Weikum[1]

[1] Max-Planck Institute for Informatics,
Saarland Informatics Campus, Saarbrücken, Germany
{htran,dstepano,gadelrab,weikum}@mpi-inf.mpg.de
[2] Università degli Studi di Bari Aldo Moro, Bari, Italy
francesca.lisi@uniba.it

Abstract. Recent advances in information extraction have led to the so-called knowledge graphs (KGs), i.e., huge collections of relational factual knowledge. Since KGs are automatically constructed, they are inherently incomplete, thus naturally treated under the Open World Assumption (OWA). Rule mining techniques have been exploited to support the crucial task of KG completion. However, these techniques can mine Horn rules, which are insufficiently expressive to capture exceptions, and might thus make incorrect predictions on missing links. Recently, a rule-based method for filling in this gap was proposed which, however, applies to a flattened representation of a KG with only unary facts. In this work we make the first steps towards extending this approach to KGs in their original relational form, and provide preliminary evaluation results on real-world KGs, which demonstrate the effectiveness of our method.

1 Introduction

Motivation. Recent advances in information extraction have led to the so-called *knowledge graphs* (KGs), *i.e.* huge collections of *triples* ⟨*subject predicate object*⟩ according to the RDF data model [17]. These triples encode facts about the world and can be straightforwardly represented by means of unary and binary first-order logic (FOL) predicates. The unary predicates are the objects of the RDF *type* predicate, while the binary ones correspond to all other RDF predicates, *e.g.*, ⟨*alice type researcher*⟩ and ⟨*bob isMarriedTo alice*⟩ from the KG in Fig. 1 refer to *researcher(alice)* and *isMarriedTo(bob, alice)* respectively. Notable examples of KGs are NELL [4], DBpedia [1], YAGO [23] and Wikidata [9].

Since KGs are automatically constructed, they are inherently *incomplete*. Therefore, they are naturally treated under the Open World Assumption (OWA). The task of *completion* (also known as *link prediction*) is of crucial importance for the curation of KGs. To this aim, rule mining techniques (*e.g.*, [5,12]) have been exploited to automatically build rules able to make predictions on missing links.

© Springer International Publishing AG 2017
J. Cussens and A. Russo (Eds.): ILP 2016, LNAI 10326, pp. 94–107, 2017.
DOI: 10.1007/978-3-319-63342-8_8

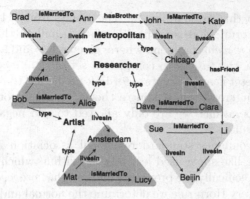

Fig. 1. Example of a knowledge graph

However, they mine Horn rules, which are insufficiently expressive to capture exceptions, and might thus deduce incorrect facts. For example, the following rule

$$r1 : livesIn(Y, Z) \leftarrow isMarriedTo(X, Y), livesIn(X, Z)$$

can be mined from the KG in Fig. 1 and used to produce the facts *livesIn(alice,berlin)*, *livesIn(dave,chicago)* and *livesIn(lucy, amsterdam)*. Observe that the first two predicted facts might actually be wrong. Indeed, both *alice* and *dave* are researchers, and the rule *r1* could be suspected to have *researcher* as a potential exception.

Challenges. Exception handling has been traditionally faced in ILP by learning *nonmonotonic logic programs*, i.e., programs with negations [6,15,18,25,26]. However, there are several important obstacles that prevent us from using the off-the-shelf nonmonotonic ILP algorithms. First, the *target predicates* can not be easily identified, since we do not know which parts of the considered KG need to be completed. A standard way of addressing this issue would be just to learn rules for all the different predicate names occurring in the KG. Unfortunately, this is unfeasible in our case given the huge size of KGs. Second, the *negative examples* are not available, and they can not be easily obtained from, *e.g.*, domain experts due to - once again - the huge size of KGs. A natural solution to cope with this issue is to learn rules from positive examples only. Third, the definition of a *language bias* turns out to be cumbersome since the schema of the KG is usually not available.

To overcome the obstacles mentioned above, it turns out to be appropriate to treat the KG completion problem as an unsupervised relational learning task, and exploit algorithms for relational association rule mining such as [12]. In [11] these techniques are applied to first learn a set of Horn rules, which subsequently can be revised by adding negated atoms to their bodies in order to account for exceptions. However, the proposed approach applies only to a flattened representation of a KG containing just unary facts.

Contributions. In this work we extend the results from [11] to KGs in their original relational form. More specifically, we reformulate the KG completion problem as a *theory revision* problem, where, given a KG and a set of (previously learned) Horn rules, the task is to compute a set of *nonmonotonic rules*, such that the revised ruleset is more accurate for link prediction than the original one. Essentially, we are interested in tackling a theory revision problem, in which, as possible revision operations, we are only allowed to add negated atoms to the antecedents of the rules.

Our approach combines standard relational association rule mining techniques with a FOIL-like supervised learning algorithm, which is used to detect exceptions. More specifically, we propose a method that proceeds in four steps as follows: First, for every Horn rule we determine the normal and abnormal substitutions, i.e., substitutions that satisfy (resp. do not satisfy) the considered rule. Second, we compute the so-called exception witness sets, i.e., sets of predicates that are potentially involved in explaining why abnormal substitutions fail to follow the rule (e.g., *researcher* in our example). Third, we construct candidate rule revisions by adding a single exception at a time. We devise quality measures for nonmonotonic rules to quantify their strength w.r.t the KG. We consider the crosstalk between the rules through the novel *partial materialization* technique instead of revising rules in isolation. Fourth, we rank rule revisions according to these measures to determine a ruleset that not only describes the data well but also shows a good predictive power by taking exceptions into account.

The contributions of our paper are:
- A theory revision framework, based on nonmonotonic relational rule learning, for capturing exceptions in rule-based approaches to KG completion.
- A methodology for computing exception candidates, measuring their quality, and ranking them taking into account the interaction among the rules.
- Experiments with the YAGO3 and IMDB KGs, which demonstrate the gains of our method for rule quality as well as fact quality when performing KG completion.

Structure. Section 2 introduces preliminaries on nonmonotonic logic programming and relational association rule mining. Section 3 describes our theory revision framework and the methodology. Section 4 reports on experimental results, while Sections 5 and 6 discuss the related work and conclude the paper respectively.

2   Preliminaries

Nonmonotonic logic programming. We consider logic programs in their usual definition [22] under the answer set semantics. In short, a *(nonmonotonic) logic program* P is a set of *rules* of the form

$$H \leftarrow B, not\ E \tag{1}$$

where H is a standard first-order atom of the form $a(\boldsymbol{X})$ known as the rule head and denoted as $head(r)$, B is a conjunction of positive atoms of the form

$b_1(\boldsymbol{Y_1}), \ldots, b_k(\boldsymbol{Y_k})$ to which we refer as $body^+(r)$, and $not\ E$, with slight abuse of notation, denotes a conjunction of atoms $not\ b_{k+1}(\boldsymbol{Y_{k+1}}), \ldots, not\ b_n(\boldsymbol{Y_n})$. Here, not is the so-called *negation as failure (NAF)* or *default negation*. The negated part of the body is denoted as $body^-(r)$. The rule r is *positive* or *Horn* if $body^-(r) = \emptyset$. $\boldsymbol{X}, \boldsymbol{Y_1}, \ldots, \boldsymbol{Y_n}$ are tuples of either constants or variables whose length corresponds to the arity of the predicates $a, b_1, \ldots, b_n$ respectively. The signature of P is given as $\Sigma_P = \langle \mathbf{P}, \mathcal{C} \rangle$, where $\mathbf{P}$ and $\mathcal{C}$ are resp. sets of predicates and constants occurring in P.

A logic program P is *ground* if it consists of only ground rules, i.e. rules without variables. Ground instantiation $Gr(P)$ of a nonground program P is obtained by substituting variables with constants in all possible ways. The *Herbrand universe* $HU(P)$ (resp. *Herbrand base* $HB(P)$) of P, is the set of all constants occurring in P, i.e. $HU(P) = \mathcal{C}$ (resp. the set of all possible ground atoms that can be formed with predicates in $\mathbf{P}$ and constants in $\mathcal{C}$). We refer to any subset of $HB(P)$ as a *Herbrand interpretation*. By $MM(P)$ we denote the set-inclusion minimal Herbrand interpretation of a ground positive program P.

An interpretation I of P is an *answer set* (or *stable model*) of P iff $I \in MM(P^I)$, where P^I is the *Gelfond-Lifschitz (GL) reduct* [13] of P, obtained from $Gr(P)$ by removing (i) each rule r such that $body^-(r) \cap I \neq \emptyset$, and (ii) all the negative atoms from the remaining rules. The set of answer sets of a program P is denoted by $AS(P)$.

Example 1. Consider the program

$$P = \left\{ \begin{array}{l} (1)\ \textit{livesIn(brad, berlin)}; \quad (2)\ \textit{isMarriedTo(brad, ann)}; \\ (3)\ \textit{livesIn(Y, Z)} \leftarrow \textit{isMarriedTo(X, Y)}, \textit{livesIn(X, Z)}, not\ \textit{researcher(Y)} \end{array} \right\}$$

The ground instantiation $Gr(P)$ of P is obtained by substituting X, Y, Z with *brad, ann* and *berlin* respectively. For $I = \{\textit{isMarriedTo(brad, ann)}, \textit{livesIn(ann, berlin)}, \textit{livesIn(brad, berlin)}\}$, the GL-reduct P^I of P contains the rule $\textit{livesIn(ann, berlin)} \leftarrow \textit{livesIn(brad, berlin)}, \textit{isMarriedTo(brad, ann)}$ and the facts (1), (2). As I is a minimal model of P^I, it holds that I is an answer set of P. □

Following the common practice in ILP, we consider only safe rules (i.e., variables in the negated part must appear in some positive atoms) with linked variables [14].

Relational association rule mining. Association rule mining concerns the discovery of frequent patterns in a data set and the subsequent transformation of these patterns into rules. Association rules in the relational format have been subject of intensive research in ILP (see, e.g., [8] as the seminal work in this direction) and more recently in the KG community (see [12] as the most prominent work). In the following we adapt basic notions in relational association rule mining to our case of interest.

A *conjunctive query* Q over $\mathcal{G}$ is of the form $Q(\boldsymbol{X}) : -p_1(\boldsymbol{X}1), \ldots, p_m(\boldsymbol{X_m})$. Its right-hand side (i.e., body) is a finite set of possibly negated atomic formulas over $\mathcal{G}$, while the left-hand side (i.e., head) is a tuple of variables occurring in the body. The *answer* of Q on $\mathcal{G}$ is the set $Q(\mathcal{G}) := \{f(\mathbf{Y}) \mid \mathbf{Y}$ is the head of Q and f is a matching of Q on $\mathcal{G}\}$. Following [8], the *(absolute) support* of a conjunctive query Q in a KG $\mathcal{G}$ is the number of distinct tuples in the answer of Q on $\mathcal{G}$. The support of the query

$$Q(X, Y, Z) : -isMarriedTo(X, Y), livesIn(X, Z) \tag{2}$$

over $\mathcal{G}$ in Fig. 1 asking for people, their spouses and living places is equal to 6.

An *association rule* is of the form $Q_1 => Q_2$, such that Q_1 and Q_2 are both conjunctive queries and the body of Q_1 considered as a set of atoms is included in the body of Q_2, i.e., $Q_1(\mathcal{G}') \subseteq Q_2(\mathcal{G}')$ for any possible KG $\mathcal{G}'$.

For example, from the above $Q(X, Y, Z)$ and

$$Q'(X, Y, Z) : -isMarriedTo(X, Y), livesIn(X, Z), livesIn(Y, Z) \tag{3}$$

we can construct the rule $Q => Q'$.

In this work we exploit association rules for reasoning purposes, and thus (with some abuse of notation) treat them as logical rules, i.e., for $Q_1 => Q_2$ we write $Q_2 \backslash Q_1 \leftarrow Q_1$, where $Q_2 \backslash Q_1$ refers to the set difference between Q_2 and Q_1 considered as sets. E.g., $Q => Q'$ from above corresponds to *r1* from Sect. 1.

We exploit the rule evaluation measure called *conviction* [3], as it is accepted to be appropriate for estimating the actual implication of the rule at hand, and is thus particularly attractive for our KG completion task. For $r : H \leftarrow B, not\ E$, with $H = h(X, Y)$ and B, E involving variables from $\mathbf{Z} \supseteq X, Y$, the *conviction* is given by:

$$conv(r, \mathcal{G}) = \frac{1 - supp(h(X, Y), \mathcal{G})}{1 - conf(r, \mathcal{G})} \tag{4}$$

where $supp(h(X, Y), \mathcal{G})$ is the *relative support* of $h(X, Y)$ defined as follows:

$$supp(h(X, Y), \mathcal{G}) = \frac{\#(X, Y) : h(X, Y) \in \mathcal{G}}{(\#X : \exists Y\ h(X, Y) \in \mathcal{G}) * (\#Y : \exists X\ h(X, Y) \in \mathcal{G})} \tag{5}$$

and *conf* is the confidence of r given as

$$conf(r, \mathcal{G}) = \frac{\#(X, Y) : H \in \mathcal{G}, \exists \mathbf{Z}\ B \in \mathcal{G}, E \notin \mathcal{G}}{\#(X, Y) : \exists \mathbf{Z}\ B \in \mathcal{G}, E \notin \mathcal{G}} \tag{6}$$

Example 2. The conviction of the rule *r1* is $conv(r1, \mathcal{G}) = \dfrac{1 - 0.3}{1 - 0.5} = 1.4$. □

3 A Theory Revision Framework for Rule-Based KG Completion

3.1 Problem Statement

We start with defining the goal of this work formally. To this aim, let us introduce the factual representation of a KG $\mathcal{G}$ as the collection of facts over the signature

$\Sigma_{\mathcal{G}} = \langle \mathbf{C}, \mathbf{R}, \mathcal{C} \rangle$, where $\mathbf{C}$, $\mathbf{R}$ and $\mathcal{C}$ are sets of unary predicates, binary predicates and constants, resp. Following [7], we define the gap between the *available graph* $\mathcal{G}^a$ and the *ideal graph* $\mathcal{G}^i$, i.e., the graph containing all correct facts with constants and relations from $\Sigma_{\mathcal{G}^a}$ that hold in the current state of the world.

Definition 1 (Incomplete data source). *An incomplete data source is a pair* $G = (\mathcal{G}^a, \mathcal{G}^i)$ *of two KGs, where* $\mathcal{G}^a \subseteq \mathcal{G}^i$ *and* $\Sigma_{\mathcal{G}^a} = \Sigma_{\mathcal{G}^i}$.

Our goal is to learn a set $\mathcal{R}$ of nonmonotonic rules from the available graph, such that their application results in a good approximation of $\mathcal{G}^i$. Here, the application of $\mathcal{R}$ to a graph $\mathcal{G}$ refers to the computation of answer sets of $\mathcal{R} \cup \mathcal{G}$.

Definition 2 (Rule-based KG completion). *Let a factual representation of a KG* $\mathcal{G}$ *be given over the signature* $\Sigma_{\mathcal{G}} = \langle \mathbf{C}, \mathbf{R}, \mathcal{C} \rangle$ *and* $\mathcal{R}$ *be a set of rules mined from* $\mathcal{G}$, *i.e. rules over the signature* $\Sigma_{\mathcal{R}} = \langle \mathbf{C} \cup \mathbf{R}, \mathcal{C} \rangle$. *Then, the* completion *of* $\mathcal{G}$ *w.r.t.* $\mathcal{R}$ *is a graph* $\mathcal{G}_{\mathcal{R}}$ *constructed from any answer set* $\mathcal{G}_{\mathcal{R}} \in AS(\mathcal{R} \cup \mathcal{G})$.

Note that $\mathcal{G}^i$ is the perfect completion of $\mathcal{G}^a$, containing all correct facts over $\Sigma_{\mathcal{G}^a}$. Given a potentially incomplete graph $\mathcal{G}^a$ and a set $\mathcal{R}_H$ of Horn rules mined from $\mathcal{G}^a$, our goal is to add default negated atoms (exceptions) to the rules in $\mathcal{R}_H$ and obtain a revised ruleset $\mathcal{R}_{NM}$ such that the set difference between $\mathcal{G}^a_{\mathcal{R}_{NM}}$ and $\mathcal{G}^i$ is as small as possible. Intuitively, a good revision $\mathcal{R}_{NM}$ of $\mathcal{R}_H$ is the one that (i) neglects as many incorrect predictions made by $\mathcal{R}_H$ as possible, while still (ii) preserving as many correct predictions made by $\mathcal{R}_H$ as possible. Note that $\mathcal{G}^i$ is usually *not available*, thus we do not know which predictions are actually correct and which are not. For this reason using standard ILP measures in our setting to evaluate the quality of a ruleset is impractical. To still make an estimate of the revision quality we exploit measures from association rule mining literature. According to our hypothesis, a good ruleset revision is the one for which the overall rule measure is the highest, while the added negated atoms are not over-fitting the data, i.e., the negated atoms are actual exceptions rather than noise.

To this end, we devise two *quality functions*, q_{rm} and $q_{conflict}$, that take a ruleset $\mathcal{R}$ and a KG $\mathcal{G}$ as input and output a real value, reflecting the suitability of $\mathcal{R}$ for data prediction. In particular, q_{rm} generalizes any rule measure rm to rulesets as follows

$$q_{rm}(\mathcal{R}, \mathcal{G}) = \frac{\sum_{r \in \mathcal{R}} rm(r, \mathcal{G})}{|\mathcal{R}|}. \tag{7}$$

Conversely, $q_{conflict}$ estimates the number of conflicting predictions that the rules in $\mathcal{R}$ generate. To measure $q_{conflict}$ for a given $\mathcal{R}$, we create an extended set of rules $\mathcal{R}^{aux}$, which contains each nonmonotonic rule $r \in \mathcal{R}$ together with its auxiliary version r^{aux}, constructed as follows: (1) transform r into a Horn rule by removing *not* from negated body atoms, and (2) replace the head predicate h of r with a newly introduced predicate not_h which intuitively contains instances which are *not* in h. Formally,

$$q_{conflict}(\mathcal{R}, \mathcal{G}) = \sum_{p \in pred(\mathcal{R})} \frac{|\{c \mid p(c), not\_p(c) \in \mathcal{G}_{\mathcal{R}^{aux}}\}|}{|\{c \mid not\_p(c) \in \mathcal{G}_{\mathcal{R}^{aux}}\}|}, \tag{8}$$

where $pred(\mathcal{R})$ is the set of predicates appearing in $\mathcal{R}$, and $\boldsymbol{c} \subseteq \mathcal{C}$ with $1 \leq |\boldsymbol{c}| \leq 2$. Note that $q_{conflict}$ is designed to distinguish real exceptions from noise, by considering the crosstalk between the rules in a set, as illustrated in the following example.

Example 3. The predicate *researcher* is a good exception for *r1* w.r.t. $\mathcal{G}$ (Fig. 1) with $bornIn(dave, chicago)$ added, i.e. it explains why for 2 out of 3 substitutions marked with red triangles the rule *r1* is not satisfied. However, this exception becomes less prominent, whenever $r2 : livesIn(X, Y) \leftarrow bornIn(X, Y), not \quad emigrant(X)$ is applied to $\mathcal{G}$. Indeed, after $livesIn$ $(dave, chicago)$ is predicted, the substitution $X/clara$, $Y/dave$, $Z/chicago$ starts satisfying *r1*, but *researcher* still holds for *dave*, which weakens the predicate *researcher* as an exception for *r1*.   □

We now define our theory revision problem based on the above quality functions.

Definition 3 (Quality-based Horn theory revision (QHTR)). *Given a set $\mathcal{R}_H$ of Horn rules over the signature Σ, a KG $\mathcal{G}$, and the quality functions q_{rm} and $q_{conflict}$, the* quality-based Horn theory revision problem *is to find a set $\mathcal{R}_{NM}$ of rules over Σ obtained by adding default negated atoms to $body(r)$ for some $r \in \mathcal{R}_H$, such that (i) $q_{rm}(\mathcal{R}_{NM}, \mathcal{G})$ is maximal, and (ii) $q_{conflict}(\mathcal{R}_{NM}, \mathcal{G})$ is minimal.*

Prior to tackling the QHTR problem we introduce the notions of r-(ab)normal substitutions and exception witness set (EWSs) that are used in our revision framework.

Definition 4 (r-(ab)normal substitutions). *Let $\mathcal{G}$ be a KG, r a Horn rule mined from $\mathcal{G}$, and let $\mathcal{V}$ be a set of variables occurring in r. Then*

- $NS(r, \mathcal{G}) = \{\theta \mid head(r)\theta, body(r)\theta \subseteq \mathcal{G}\}$ *is an r-normal set of substitutions;*
- $ABS(r, \mathcal{G}) = \{\theta' \mid body(r)\theta' \subseteq \mathcal{G}, head(r)\theta' \notin \mathcal{G}\}$ *is an r-abnormal one, where $\theta, \theta' : \mathcal{V} \rightarrow \mathcal{C}$.*

Example 4. For $\mathcal{G}$ from Fig. 1 and *r1* we have $NS(r1, \mathcal{G}) = \{\theta_1, \theta_2, \theta_3\}$, where $\theta_1 = \{X/Brad, Y/Ann, Z/Berlin\}$; similarly, the most right and bottom blue triangles in Fig. 1 refer to θ_2 and θ_3 resp., while the red ones represent $ABS(r1, \mathcal{G})$.

Intuitively, if the given data was complete, then the r-normal and r-abnormal substitutions would exactly correspond to substitutions for which the rule r holds (resp. does not hold) in $\mathcal{G}^i$. However, some r-abnormal substitutions might be classified as such due to the OWA. In order to distinguish the "wrongly" and "correctly" classified substitutions in the r-abnormal set, we construct *exception witness sets (EWS)*.

Definition 5 Exception witness set (EWS). *Let $\mathcal{G}$ be a KG, let r be a rule mined from it, let $\mathcal{V}$ be a set of variables occurring in r and $\boldsymbol{X} \subseteq \mathcal{V}$. Exception witness set for r w.r.t. $\mathcal{G}$ and $\boldsymbol{X}$ is a maximal set of predicates $EWS(r, \mathcal{G}, \boldsymbol{X}) = \{e_1, \ldots, e_k\}$, s.t.*

- $e_i(\boldsymbol{X}\theta_j) \in \mathcal{G}$ for some $\theta_j \in ABS(r, \mathcal{G})$, $1 \le i \le k$ and
- $e_1(\boldsymbol{X}\theta'), \ldots, e_k(\boldsymbol{X}\theta') \notin \mathcal{G}$ for all $\theta' \in NS(r, \mathcal{G})$.

Example 5. For $\mathcal{G}$ in Fig. 1 and $r1$ we have that $EWS(r, \mathcal{G}, Y) = \{researcher\}$. Furthermore, $EWS(r, \mathcal{G}, X) = \{artist\}$. If *brad* with *ann* and *john* with *kate* lived in cities different from *berlin* and *chicago* resp., then $EWS(r, \mathcal{G}, Z) = \{metropolitan\}$.

In general when binary atoms are allowed in the rules, there might be potentially too many possible *EWSs* to construct. For a rule with n distinct variables, n^2 candidate *EWSs* might exist. Furthermore, combinations of exception candidates could be an explanation for some missing links, so the search space of solutions to QHTR problem is large. In this work, however, we restrict ourselves only to a single predicate as a final exception, and leave the extensions to arbitrary combinations for future research.

3.2   Methodology

Due to the large number of exception candidates to consider, determining the globally best solution to the QHTR problem is not feasible in practice especially given the huge size of KGs. Therefore, we aim at finding an approximately good solution. Intuitively, our approach is to revise rules one by one finding the locally best revision, while considering the predictive impact of other rules in a set. Our methodology for solving the QHTR problem comprises four steps, which we now discuss in details.

Step 1. We start with a KG $\mathcal{G}$ and compute frequent conjunctive queries, which are then cast into Horn rules $\mathcal{R}_H$ based on some association rule measure rm. For that any state-of-the-art relational association rule learning algorithm can be used. We then compute for each rule $r \in \mathcal{R}_H$ the *r-normal* and *r-abnormal* substitutions.

Steps 2 and 3. Then, for every $r \in \mathcal{R}_H$ with $h(X, Y)$ in the head, we determine $EWS(r, \mathcal{G}, X)$, $EWS(r, \mathcal{G}, Y)$ and $EWS(r, \mathcal{G}, \langle X, Y \rangle)$. The algorithm for computing *EWSs* is an extended version of the one reported in [11]. Here, we first construct $E^+ = \{not\_h(c, d), \text{s.t. } \theta = \{X/c, Y/d, \ldots\}$ is in $ABS(r, \mathcal{G})\}$ and $E^- = \{not\_h(e, f), \text{s.t. } \theta' = \{X/e, Y/f, \ldots\}$ is in $NS(r, \mathcal{G})\}$. A classical ILP procedure $learn(E^+, E^-, \mathcal{G})$ (e.g., based on [28]) is then invoked, which searches for hypothesis with $not\_h(X, Y)$ in the head and a single body atom of the form $p(X)$, $p'(Y)$ or $p''(X, Y)$, where p, p', p'' are predicates from $\Sigma_\mathcal{G}$. The target hypothesis should not cover any examples in E^-, while covering at least some examples in E^+. From the bodies of the obtained hypothesis the predicates for *EWS* sets are extracted.

   Then, for every $r \in \mathcal{R}_H$ we create potential revisions by adding to r a single negated atom from *EWS* sets at a time. Overall for each rule this way we obtain $|EWS(r, \mathcal{G}, X)| + |EWS(r, \mathcal{G}, Y)| + |EWS(r, \mathcal{G}, \langle X, Y \rangle)|$ candidate revisions.

Steps 4. After all potential revisions are constructed, we rank them and determine the resulting set $\mathcal{R}_{NM}$ by selecting for every rule the revision that is ranked the highest. To find such globally best revised ruleset $\mathcal{R}_{NM}$, too many candidate combinations have to be checked, which is impractical due to the large size of both $\mathcal{G}$ and $EWSs$. Thus, instead we incrementally build $\mathcal{R}_{NM}$ by considering every $r_i \in \mathcal{R}_H$ and choosing the locally best revision r_i^j for it. For that, we exploit three ranking functions: a naive one and two more sophisticated ones, which invoke the novel concept of *partial materialization* (**PM**). Intuitively, the idea behind it is to rank candidate revisions not based on $\mathcal{G}$, but rather on its extension with predictions produced by other, selectively chosen, rules (grouped into a set $\mathcal{R}'$), thus ensuring a crosstalk between the rules. We now describe the ranking functions in more details.

The **Naive (N)** ranker is the simplest function, which prefers the revision r_i^j with the highest value of $rm(r_i^j, \mathcal{G})$ among all revisions of r_i. This selection function produces a globally best revision with respect to (i) in Definition 3. However, it completely ignores (ii), and thus might return rules with overly noisy exceptions.

The **PM** ranker prefers r_i^j with the highest value of

$$score(r_i^j, \mathcal{G}) = \frac{rm(r_i^j, \mathcal{G}_{\mathcal{R}'}) + rm(r_i^{j\,aux}, \mathcal{G}_{\mathcal{R}'})}{2} \tag{9}$$

where $\mathcal{R}'$ is the set of rules $r_l' \in \mathcal{R}_H \backslash r_i$ with candidate exceptions from all $EWSs$ for r_l incorporated at once. Informally, $\mathcal{G}_{\mathcal{R}'}$ contains only facts that can be safely predicted by the rules from $\mathcal{R}_H \backslash r_i$, i.e., there is no evident reason (candidate exceptions) for not making these predictions, and thus we can rely on them when revising r_i.

The **OPM** ranker is similar to **PM**, but the selected ruleset $\mathcal{R}'$ contains only those rules whose Horn version appears above the considered rule r_i in the ruleset $\mathcal{R}_H$, ordered (**O**) based on some rule measure, which is not necessarily the same as rm.

4   Evaluation

Our revision approach aims at (1) enhancing the quality of a given ruleset w.r.t. conviction, and consequently (2) improving the accuracy of its predictions. Ideally, the set difference between $\mathcal{G}_{\mathcal{R}_{NM}}$ and $\mathcal{G}^i$ should be minimized (see Fig. 2 for illustration).

Dataset. An automatic evaluation of the prediction quality requires an ideal graph $\mathcal{G}^i$ which is known to be complete as a ground truth. However, obtaining a real life complete KG is not possible. Therefore, we used the existing KG as an approximation of $\mathcal{G}^i$ ($\mathcal{G}^i_{appr}$), and constructed the available graph $\mathcal{G}^a$ by removing from $\mathcal{G}^i_{appr}$ 20% of the facts for each binary predicate. As a side constraint, we ensure that every node in $\mathcal{G}^a$ is connected to at least one other node. We constructed two datasets for evaluating our approach: (i) YAGO3 [23], as a

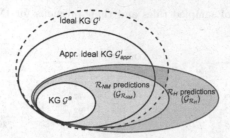

Fig. 2. Relations between the ideal, approximated and available slices of a KG.

Table 1. The average conviction for the *top-k* Horn rules and their revisions.

topk	YAGO				IMDB			
	$\mathcal{R}_H$	$\mathcal{R}_N$	$\mathcal{R}_{PM}$	$\mathcal{R}_{OPM}$	$\mathcal{R}_H$	$\mathcal{R}_N$	$\mathcal{R}_{PM}$	$\mathcal{R}_{OPM}$
5	1.3784	1.3821	1.3821	1.3821	2.2670	2.3014	2.3008	2.3014
30	1.1207	1.1253	1.1236	1.1237	1.5453	1.5644	1.5543	1.5640
50	1.0884	1.0923	1.0909	1.0913	1.3571	1.3749	1.3666	1.3746
60	1.0797	1.0837	1.0823	1.0829	1.3063	1.3221	1.3143	1.3219
70	1.0714	1.0755	1.0736	1.0744	1.2675	1.2817	1.2746	1.2814
80	1.0685	1.0731	1.0710	1.0720	1.2368	1.2499	1.2431	1.2497
100	1.0618	1.0668	1.0648	1.0659	1.3074	1.4100	1.3987	1.4098

general purpose KG, with more than 1.8M entities, 38 relations, and 20.7M facts, and (ii) a domain-specific KG extracted from the IMDB[1] dataset with 112 K entities, 38 relations, and 583 K facts[2].

Setup. We have implemented our approach in a system prototype[3], and conducted experiments on a multi-core Linux server with 40 cores and 400GB RAM. We start with mining Horn rules of the form $h(X, Z) \leftarrow p(X, Y), q(Y, Z)$ from $\mathcal{G}^a$ and ranking them w.r.t. their *absolute support*. Then, we revise the rules as described in Sect. 3.2, taking *conviction* as the *rm* measure. For every rule we rank the constructed revisions and pick the one with the highest score as the final result. This process is repeated for the proposed ranking methods, i.e., *Naive, Partial Materialization,* and *Ordered Partial Materialization* resulting in the rulesets $\mathcal{R}_N$, $\mathcal{R}_{PM}$, and $\mathcal{R}_{OPM}$ respectively.

Ruleset quality. In Table 1, we report the *average conviction* for the top-*k* (*k*=5,...100) Horn rules $\mathcal{R}_H$ and their revisions for YAGO and IMDB. The results show that the revision process consistently enhances the avg. ruleset conviction. Moreover, while the conviction per ruleset naturally decreases with addition of

---

[1] http://imdb.com.
[2] http://people.mpi-inf.mpg.de/~gadelrab/downloads/ILP2016.
[3] https://github.com/htran010589/nonmonotonic-rule-mining.

Table 2. Predictions of sampled rules and their revisions for IMDB (I) and YAGO (Y).

Predicate	Predictions				Outside $\mathcal{G}^i_{appr}$				Corr. removed, %		
	$\mathcal{R}_H$	$\mathcal{R}_N$	$\mathcal{R}_{PM}$	$\mathcal{R}_{OPM}$	$\mathcal{R}_H$	$\mathcal{R}_N$	$\mathcal{R}_{PM}$	$\mathcal{R}_{OPM}$	$\mathcal{R}_N$	$\mathcal{R}_{PM}$	$\mathcal{R}_{OPM}$
$I{:}actedIn$	1231	1214	1230	1214	1148	1131	1147	1131	90	100	90
$I{:}genre$	629	609	618	609	493	477	482	477	50	20	50
$I{:}hasLang$	173	102	125	102	163	92	115	92	60	100	60
$I{:}prodIn$	2489	2256	2327	2327	2488	2255	2326	2326	10	10	30
									52.50	45.16	**57.75**
$Y{:}direct$	41079	39174	39174	39174	41021	39116	39116	39116	100	100	100
$Y{:}grFrom$	3519	3456	3456	3456	3363	3300	3300	3300	100	100	70
$Y{:}citizOf$	3407	2883	2883	2883	3360	2836	2836	2836	50	50	70
$Y{:}bornIn$	110283	108317	109846	108317	109572	107607	109137	107607	90	90	100
									85	85	85

lower quality rules, improvement ratios are increasing with the best enhancement (7.6%) for IMDB top-100 rules.

Prediction quality. To evaluate the quality of ruleset predictions, we sampled a set of 5 Horn rules $\mathcal{R}_H$ from the top-50 Horn rules both for IMDB and YAGO and compared them against their revisions w.r.t. the predictive power. For that, we run DLV [20] with these rulesets and the facts in $\mathcal{G}^a$ and obtained resp. $\mathcal{G}_{\mathcal{R}_H}$, $\mathcal{G}_{\mathcal{R}_N}$, $\mathcal{G}_{\mathcal{R}_{PM}}$ and $\mathcal{G}_{\mathcal{R}_{OPM}}$. Table 2 reports for each head predicate appearing in the sampled rules the number of newly predicted facts, i.e. those not in $\mathcal{G}^a$ (second column) and the portion of predictions among them that are outside $\mathcal{G}^i_{appr}$ (third column).

First, observe that naturally relatively few predictions can be found in $\mathcal{G}^i_{appr}$ ($\approx$9% for IMDB and $\approx$2% for YAGO). This is expected as the latter graph is highly incomplete. Second, it is important to note that $\mathcal{R}_H$ and the revised rulesets produced roughly the same number of correct predictions within $\mathcal{G}^i_{appr}$. E.g., for YAGO we have $\mathcal{G}_{\mathcal{R}_H} \backslash \mathcal{G}_{\mathcal{R}_{PM}} \cap \mathcal{G}^i_{appr} = \emptyset$, meaning that the green area within the approximation of the ideal graph in Fig. 2 is empty, which shows that incorporated exceptions did not spoil the positive rules with respect to correct predictions in $\mathcal{G}^i_{appr}$.

To make the comparison between $\mathcal{R}_H$ and the revised rulesets fair, we need to ensure that $\mathcal{R}_H$ on its own is not completely inaccurate. Indeed, if $\mathcal{R}_H$ makes only false predictions, then adding even irrelevant exceptions will reduce the number of incorrect instances, thus, improving the ruleset predictive quality. The number of $\mathcal{R}_H$ predictions outside $\mathcal{G}^i_{appr}$ is large, and we do not know the ground truth for these predictions. Therefore, we had to verify these facts manually using web resources. Obviously such verification for all of the predictions is not feasible. Hence, we restricted ourselves to a uniform random sample of 20 predicted facts per head predicate in $\mathcal{R}_H$. Among the IMDB samples, the precision of 70% has been achieved, while for YAGO we have obtained precision of 30%. This shows that the rules in $\mathcal{R}_H$ are not completely erroneous.

To assess the impact of the revision methods, we also had to select a uniform sample due to the large size of the differences between $\mathcal{G}_{\mathcal{R}_H}$ and the graphs obtained by applying revised rulesets. More specifically, we have randomly sampled 10 predictions per head predicate from $\mathcal{G}_{\mathcal{R}_H} \backslash \mathcal{G}_{\mathcal{R}_N}$, $\mathcal{G}_{\mathcal{R}_H} \backslash \mathcal{G}_{\mathcal{R}_{PM}}$ and $\mathcal{G}_{\mathcal{R}_H} \backslash \mathcal{G}_{\mathcal{R}_{OPM}}$ resp. The 4th column in Table 2 reports the percentage of erroneous predictions among the sampled facts in the difference for each revision method (referred to as correctly removed), i.e., gray area in Fig. 2. For IMDB $\mathcal{R}_{OPM}$ achieved the best improvement. For YAGO, all of the revision methods performed equally well. Moreover, the effect of YAGO revisions is more visible, since $\mathcal{R}_H$ for YAGO is of a lower quality than for IMDB as reported earlier.

Running times. Our main goal was to evaluate the predictive quality of computed rules rather then the running times of the implemented algorithms. Therefore, the latter are only briefly reported. For the *top-100* Horn YAGO and IMDB rules mined from $\mathcal{G}^a$, *EWS*s with an average of 1.6 K and 10.9 K exception candidates per rule were computed within 7 and 68 s resp. As regards IMDB, the revisions $\mathcal{R}_N$, $\mathcal{R}_{PM}$, and $\mathcal{R}_{OPM}$ were determined in 9, 62, and 24 s resp., while for YAGO, they required 45, 177, and 112 s. Besides, the predictions of each of the rulesets on $\mathcal{G}^a$ were found via DLV, on average, within 8 s for IMDB and 310 s for YAGO.

Example rules. Figure 3 shows examples of our revised rules, e.g., r_1 extracted from IMDB states that movie plot writers stay the same throughout the sequel unless a movie is American, and r_3 learned from YAGO says that ancestors of politicians are also politicians in the same country with the exception of Mexican vice-presidents.

$r_1 : writtenBy(X, Z) \leftarrow hasPredecessor(X, Y), writtenBy(Y, Z), \textbf{not } american\_film(X)$

$r_2 : actedIn(X, Z) \leftarrow isMarriedTo(X, Y), directed(Y, Z), \textbf{not } silent\_film\_actor(X)$

$r_3 : isPoliticianOf(X, Z) \leftarrow hasChild(X, Y), isPoliticianOf(Y, Z), \textbf{not } vicepresidentOfMexico(X)$

Fig. 3. Examples of the revised rules

5   Related Work

Approaches for link prediction are divided into statistics-based (see [24] for overview), and logic-based (e.g., [12]), which are the closest to our work. The latter basically extend and adapt previous work in ILP on relational association rule mining. However, algorithms such as [12] mine only Horn rules, rather than nonmonotonic as we do.

In the association rule mining community, some works studied (interesting) exception rules (e.g. [27]), i.e., rules with low support and high confidence. Our work differs as we do not necessarily look for rare rules, but care about their predictive power.

In the context of inductive and abductive logic [10], learning nonmonotonic rules from complete datasets was considered in several works [6, 18, 25, 26]

These methods rely on CWA and focus on describing a dataset at hand exploiting negative examples, which are explicitly given unlike in our setting. Learning nonmonotonic rules in presence of incompleteness was studied in hybrid settings in [16,21] respectively. There a background theory or a hypothesis can be represented as a combination of a DL ontology and Horn or nonmonotonic rules. While the focus of these works is on the complex interaction between reasoning components, we are more concerned with techniques for deriving rules with high predictive quality from huge KGs.

6   Conclusions and Future Work

We have presented an approach for mining relational nonmonotonic rules from KGs under OWA by casting this problem into a theory revision task and exploiting association rule mining methods to cope with the huge size of KGs. The approach extends our previous work [11], where this problem was studied for KGs with only unary predicates.

Further extensions to more complex combinations of exceptions as well as more general types of rules (e.g., with existentials in the head) are a natural future direction. Moreover enhancing our framework by partial completeness assumptions for certain (combinations of) predicates/constants is another orthogonal but interesting research stream. On the practical side, we plan to develop advanced evaluation strategies, which is very challenging due to the absence of the ideal graph and the large KG size.

Acknowledgements. We thank anonymous reviewers for their insightful suggestions and Jacopo Urbani for his helpful comments on an earlier version of this paper.

References

1. Auer, S., Bizer, C., Kobilarov, G., Lehmann, J., Cyganiak, R., Ives, Z.: DBpedia: a nucleus for a web of open data. In: Aberer, K., et al. (eds.) ASWC/ISWC - 2007. LNCS, vol. 4825, pp. 722–735. Springer, Heidelberg (2007). doi:10.1007/978-3-540-76298-0_52
2. Azevedo, P.J., Jorge, A.M.: Comparing rule measures for predictive association rules. In: ECML, pp. 510–517 (2007)
3. Brin, S., Motwani, R., Ullman, J.D., Tsur, S.: Dynamic itemset counting and implication rules for market basket data. In: SIGMOD, pp. 255–264 (1997)
4. Carlson, A., Betteridge, J., Kisiel, B., Settles, B., Jr., E.R.H., Mitchell, T.M.: Toward an architecture for never-ending language learning. In: AAAI (2010)
5. Chen, Y., Goldberg, S., Wang, D.Z., Johri, S.S.: Ontological pathfinding: mining first-order knowledge from large knowledge bases. In: SIGMOD (2016)
6. Corapi, D., Russo, A., Lupu, E.: Inductive logic programming as abductive search. In: ICLP, pp. 54–63 (2010)
7. Darari, F., Nutt, W., Pirrò, G., Razniewski, S.: Completeness statements about RDF data sources and their use for query answering. In: Alani, H., et al. (eds.) ISWC 2013. LNCS, vol. 8218, pp. 66–83. Springer, Heidelberg (2013). doi:10.1007/978-3-642-41335-3_5

8. Dehaspe, L., Raedt, L.: Mining association rules in multiple relations. In: Lavrač, N., Džeroski, S. (eds.) ILP 1997. LNCS, vol. 1297, pp. 125–132. Springer, Heidelberg (1997). doi:10.1007/3540635149_40

9. Erxleben, F., Günther, M., Krötzsch, M., Mendez, J., Vrandečić, D.: Introducing wikidata to the linked data web. In: Mika, P., et al. (eds.) ISWC 2014. LNCS, vol. 8796, pp. 50–65. Springer, Cham (2014). doi:10.1007/978-3-319-11964-9_4

10. Flach, P.A., Kakas, A.: Abduction and Induction: Essays on Their Relation and Integration, vol. 18. Applied Logic Series (2000)

11. Gad-Elrab, M.H., Stepanova, D., Urbani, J., Weikum, G.: Exception-enriched rule learning from knowledge graphs. In: Groth, P., et al. (eds.) ISWC 2016. LNCS, vol. 9981, pp. 234–251. Springer, Cham (2016). doi:10.1007/978-3-319-46523-4_15

12. Galárraga, L., Teflioudi, C., Hose, K., Suchanek, F.M.: Fast rule mining in ontological knowledge bases with AMIE+. In: VLDB J. (2015)

13. Gelfond, M., Lifschitz, V.: The stable model semantics for logic programming. In: ICLP, pp. 1070–1080 (1988)

14. Helft, N.: Induction as nonmonotonic inference. In: KR, pp. 149–156 (1989)

15. Inoue, K., Kudoh, Y.: Learning extended logic programs. In: IJCAI, pp. 176–181 (1997)

16. Józefowska, J., Lawrynowicz, A., Lukaszewski, T.: The role of semantics in mining frequent patterns from knowledge bases in description logics with rules. TPLP 10(3), 251–289 (2010)

17. Lassila, O., Swick, R.R.: Resource description framework (RDF) model and syntax specification (1999)

18. Law, M., Russo, A., Broda, K.: Inductive learning of answer set programs. In: Fermé, E., Leite, J. (eds.) JELIA 2014. LNCS, vol. 8761, pp. 311–325. Springer, Cham (2014). doi:10.1007/978-3-319-11558-0_22

19. Lehmann, J., Auer, S., Bühmann, L., Tramp, S.: Class expression learning for ontology engineering. J. Web Sem. 9(1), 71–81 (2011)

20. Leone, N., Pfeifer, G., Faber, W., Eiter, T., Gottlob, G., Perri, S., Scarcello, F.: The dlv system for knowledge representation and reasoning. ACM TOCL 7(3), 499–562 (2006)

21. Lisi, F.A.: Inductive logic programming in databases: from datalog to DL+log. TPLP 10(3), 331–359 (2010)

22. Lloyd, J.W.: Foundations of Logic Programming, 2nd edn. Springer, Berlin Heidelberg (1987)

23. Mahdisoltani, F., Biega, J., Suchanek, F.M.: YAGO3: A knowledge base from multilingual wikipedias. In: Proceedings of CIDR (2015)

24. Nickel, M., Murphy, K., Tresp, V., Gabrilovich, E.: A review of relational machine learning for knowledge graphs. Proc. IEEE 104(1), 11–33 (2016)

25. Ray, O.: Nonmonotonic abductive inductive learning. J. Appl. Log. 3(7), 329–340 (2008)

26. Sakama, C.: Induction from answer sets in nonmonotonic logic programs. ACM Trans. Comput. Log. 6(2), 203–231 (2005)

27. Taniar, D., Rahayu, W., Lee, V., Daly, O.: Exception rules in association rule mining. Appl. Math. Comput. 205(2), 735–750 (2008)

28. Quinlan, J.R.: Learning logical definitions from relations. Mach. Learn. 5, 239–266 (1990)

29. Wrobel, S.: First order theory refinement. In: ILP, pp. 14–33 (1996)

Learning Predictive Categories Using Lifted Relational Neural Networks

Gustav Šourek[1]([✉]), Suresh Manandhar[2], Filip Železný[1], Steven Schockaert[3], and Ondřej Kuželka[3]

[1] Czech Technical University, Prague, Czech Republic
{souregus,zelezny}@fel.cvut.cz
[2] Department of Computer Science, University of York, York, UK
suresh.manandhar@york.ac.uk
[3] School of CS and Informatics, Cardiff University, Cardiff, UK
{SchockaertS1,Kuzelka0}@cardiff.ac.uk

Abstract. Lifted relational neural networks (LRNNs) are a flexible neural-symbolic framework based on the idea of lifted modelling. In this paper we show how LRNNs can be easily used to specify declaratively and solve learning problems in which latent categories of entities, properties and relations need to be jointly induced.

1 Introduction

Lifted models, such as Markov logic networks (MLNs [13]), are first-order representations that define patterns from which specific (ground) models can be unfolded. For example, in a MLN we may express the pattern that *friends of smokers tend to be smokers*, which then constrains the probabilistic relationships between specific individuals in the derived ground Markov network. Inspired by this idea, in [16] we introduced a method that uses weighted relational rules for learning feed-forward neural networks, called *Lifted Relational Neural Networks* (LRNNs). This approach differs from standard neural networks in two important ways: (i) the network structure is derived from symbolic rules and thus has an intuitive interpretation, and (ii) the weights of the network are tied to the first-order rules and are thus shared among different neurons.

In this paper, we first show how LRNNs can be used to learn a latent category structure that is predictive in the sense that the properties of a given entity can be largely determined by the category to which that entity belongs, and dually, the entities satisfying a given property can be largely determined by the category to which that property belongs. This enables a form of transductive reasoning which is based on the idea that similar entities have similar properties. We then extend this model into a relational setting, in which entities not only have properties but can also be linked by arbitrary relations.

The proposed approach is similar in spirit to [7], which instead uses crisp clustering based on second-order MLNs. However, the use of LRNNs has several important advantages for learning latent concepts. Firstly, LRNNs do not need

© Springer International Publishing AG 2017
J. Cussens and A. Russo (Eds.): ILP 2016, LNAI 10326, pp. 108–119, 2017.
DOI: 10.1007/978-3-319-63342-8_9

to invoke costly EM algorithms and hence can be more efficient than latent variable probabilistic models. Secondly, the learnt soft clusters can naturally be interpreted as vector space embeddings of entities, properties and relations. Finally, the flexibility of LRNNs means that the considered form of transductive reasoning can be extended in a natural way to take into account various forms of prior domain knowledge, as well as alternative types of heuristic reasoning (e.g. reasoning by analogy, modelling persistence or periodic behaviour).

The remainder of this paper is structured as follows. In Sect. 2 we briefly describe the LRNN framework from [16], after which we introduce a technique to deal with recursive rules in LRNNs. We describe the predictive model for the non-relational setting in Sect. 3.1 and for the relational setting in Sect. 3.2. Next, in Sect. 4, we describe a simple model encoded as a LRNN which is based on similarity-based reasoning. In Sect. 5 we evaluate the method experimentally. Finally, we discuss related work in Sect. 6 and conclude the paper in Sect. 7

2 Lifted Relational Neural Networks

2.1 The Basic Framework

A lifted relational neural network (LRNN) $\mathcal{N}$ is a set of weighted definite first-order clauses[1]. Let $\mathcal{H}_{\mathcal{N}}$ be the least Herbrand model of the classical theory $\{\alpha : (\alpha, w) \in \mathcal{N}\}$, with $\mathcal{N}$ a LRNN. We define the *grounding of* $\mathcal{N}$ as $\overline{N} = \{(h\theta \leftarrow b_1\theta \wedge \cdots \wedge b_k\theta, w) : (h \leftarrow b_1 \wedge \cdots \wedge b_k, w) \in \mathcal{N}$ and $\{h\theta, b_1\theta, \ldots, b_k\theta\} \subseteq \mathcal{H}_{\mathcal{N}}\}$.

Definition 1. *Let $\mathcal{N}$ be a LRNN, and let $\overline{N}$ be its grounding. Let $g_{\vee}$, $g_{\wedge}$ and g_* be functions[2] from $\bigcup_{i=1}^{\infty} \mathbb{R}^i$ to $\mathbb{R}$. The ground neural network of $\mathcal{N}$ is a feed-forward neural network constructed as follows.*

- *For every ground atom h occurring in $\overline{N}$, there is a neuron A_h with activation function $g_{\vee}$, called* atom neuron.
- *For every ground fact $(h, w) \in \overline{N}$, there is a neuron $F_{(h,w)}$, called* fact neuron, *which has no input and always outputs the constant value w.*
- *For every ground rule $(h\theta \leftarrow b_1\theta \wedge \cdots \wedge b_k\theta, w) \in \overline{N}$, there is a neuron $R_{h\theta \leftarrow b_1\theta \wedge \cdots \wedge b_k\theta}$ with activation function $g_{\wedge}$, called* rule neuron. *It has the atom neurons $A_{b_1\theta}, \ldots, A_{b_k\theta}$ as inputs, all with weight 1.*
- *For every rule $(h \leftarrow b_1 \wedge \cdots \wedge b_k, w) \in \mathcal{N}$ and every $h\theta \in \mathcal{H}_{\mathcal{N}}$, there is a neuron $Agg_{(h \leftarrow b_1 \wedge \cdots \wedge b_k, w)}^{h\theta}$ with activation function g_*, called* aggregation neuron. *Its inputs are all rule neurons $R_{h\theta' \leftarrow b_1\theta' \wedge \cdots \wedge b_k\theta'}$ where $h\theta = h\theta'$ with all weights equal to 1.*
- *Inputs of an atom neuron $A_{h\theta}$ are the aggregation neurons $Agg_{(h \leftarrow b_1 \wedge \cdots \wedge b_k, w)}^{h\theta}$ and fact neurons $F_{(h\theta, w)}$, with the input weights determined by the outputs of the aggregation and fact neurons.*

---

[1] Established notions such as "rule" are further used also for their weighted analogies.

[2] These represent aggregation operators that can take a variable number of arguments.

Depending on the used families of activation functions $g_\wedge$, $g_\vee$ and g_*, we can obtain neural networks with different behavior. In this paper we will use:

$$g_\wedge(b_1, \ldots, b_k) = sigm\Big(\sum_{i=1}^{k} b_i - k + b_0 \Big) \quad g_\vee(b_1, \ldots, b_k) = sigm\Big(\sum_{i=1}^{k} b_i + b_0 \Big)$$

$$g_*(b_1, \ldots, b_m) = \frac{1}{m} \sum_{i=1}^{m} b_i$$

Where $sigm$ denotes the logistic sigmoid function $sigm(x) = \frac{1}{1+(e^{-x})}$, which also implies that $\forall i : 0 < b_i < 1$. Note that $g_\wedge$ and $g_\vee$ are closely related to the conjunction and disjunction from Łukasiewicz logic [6], which is in accordance with the intuition that $g_\wedge$ should only have a high output if all its inputs are high, while $g_\vee$ should be high as soon as one of the inputs is high.

2.2 Handling Recursion in LRNNs

As originally introduced in [16], LRNNs did not support recursive rules, in order to avoid the potential need to work with recurrent neural networks, since these are more difficult to train than feed-forward neural networks. However in general, recursive rules do not necessarily pose a problem to the LRNN framework as long as they do not induce directed cycles in the resulting ground neural networks. For instance, rules defining directed paths in acyclic graphs would not lead to directed cycles in the resulting ground neural networks, despite being recursive. One minor complication caused by allowing LRNNs to have recursion, even in the absence of directed cycles, is that weights may be shared among neurons that lie on a directed path from the input to the output of the network. This makes the computation of gradients more complicated than in the normal case. Although weights in such LRNNs, whose groundings are still feed-forward neural networks, can still be learned using Stochastic Gradient Descent (SGD).

In this paper we will use recursive rule sets that may potentially lead to recurrent neural networks. In order to maintain the feed-forward nature of the resulting ground neural networks, we modify the strategy for constructing ground networks as follows. First we construct the ground network exactly as described in Sect. 2. If this network contains directed cycles, we then proceed as follows. Let Q be a given ground query atom[3]. We find the respective atom neuron corresponding to Q in the ground network. If no such atom neuron exists, the output value for Q is 0. If there is such an atom neuron, we perform a breadth-first search from this atom neuron (traversing the connections between neurons in reverse, i.e. from output to input) and whenever we find an edge pointing *from* an already visited atom neuron, we delete it. The resulting ground neural network is then feed-forward. While this process enables us to stick with feed-forward neural networks, it comes at the price of a slightly less intuitive semantics, in

---

[3] In general LRNNs support non-ground query atoms but in this paper we will not need them. Therefore we assume only ground query atoms for simplicity.

which the inference and output for non-query atom neurons may also depend on the used queries. This is not problematic for any of the applications considered in this paper, as non-query atoms are not used within these application.

3 Learning Predictive Categories

In this section, we introduce a class of LRNN models that are aimed at learning predictive categories of entities, properties and relations. We first introduce a model for attribute-valued data in Sect. 3.1, which is extended to cope with relational data in Sect. 3.2.

3.1 Predictive Categories for Attribute-Valued Data

Let a set of entities be given, and for each entity, a list of properties that it satisfies. The basic assumption underlying our model is that there exist some (possibly overlapping) categories, such that every entity can be described accurately enough by its soft membership to each of these categories. We furthermore assume that these categories can themselves be organised in a set of higher-level categories. The idea is that the category hierarchy should allow us to predict which properties a given entity has, where the properties associated with higher-level categories are typically (but not necessarily) inherited by their sub-categories. To improve the generalization ability of our method, we assume that a dual category structure exists for properties. The main task we consider is to learn these (latent) category structures from the given input data.

To encode the above described model in a LRNN, we proceed as follows. We use $HasProperty(e, p)$ to denote that the entity e has the property p. For every entity e and for each category c at the lowest level of the category hierarchy, we construct the following ground rule:

$$w_{ec} : IsA(e, c)$$

Note that weight w_{ec} intuitively reflects the soft membership of e to the category c; it will be determined when training the ground network. Similarly, for each category c_1 at a given level and each category c_2 one level above, we add the following ground rule:

$$w_{c_1 c_2} : IsA(c_1, c_2)$$

In the same way, ground rules are added that link each property to a property category at the lowest level, as well as ground rules that link property categories to higher-level categories. To encode the idea that entity categories should be predictive of properties, we add the following rule for each entity category c_e and each property category c_p:

$$w_{c_e c_p} : HasProperty(A, B) \leftarrow IsA(A, c_e), IsA(B, c_p).$$

The weights $w_{c_e c_p}$ encode which entity categories are related to which property categories, and will again be determined when training weights of the LRNN. To encode transitivity of the is-a relationship, we simply add the following rule:

$$w_{isa} : IsA(A, C) \leftarrow IsA(A, B), IsA(B, C).$$

Training examples are encoded as a set of facts of the form $(HasProperty(e, p), l)$ where $l \in \{0, 1\}$, 0 denoting a negative example and 1 a positive example. We train the model using SGD as described in [16]. In particular, in a LRNN, there is a neuron for any ground literal which is logically entailed by the rules and facts in the LRNN and the output of this neuron represents the truth value of this literal. Therefore if we want to train the weights of the LRNN, we just optimize the weights of the network w.r.t. a loss function such as the mean squared error, where the loss function is computed from the desired truth values of the query literals and the outputs obtained from the respective atom neurons.

3.2   Predictive Categories for Relational Data

The model from Sect. 3.1 can be extended to cope with relational facts. Similar to our encoding of properties, we will use a reified representation of relational facts, writing e.g. $Relation(ParentOf, e_1, e_2)$ to denote that e_1 is the parent of e_2. In this way, we can induce predictive relation categories, similar to the entity and property categories considered in Section 3.1.

To this end, analogously as for entity and property categories, for every relation r and every (latent) relation category c we add the following ground rule:

$$w_{rc} : IsA(r, c)$$

For each relation category c_1 at a given level and each category c_2 one level above, we add the following ground rule:

$$w_{c_1 c_2} : IsA(c_1, c_2).$$

Note that a rule encoding transitivity of the IsA relation was already added in the first part of the model. Finally we encode that, like properties, relations among entities are typically determined by their categories. Specifically, for each triple consisting of a pair of (not necessarily distinct) entity clusters c_e, c'_e and a relation cluster c_r, we add the following ground rule:

$$w_{c_r c_e c'_e} : Relation(R, A, B) \leftarrow IsA(R, c_r), IsA(A, c_e), IsA(B, c'_e) \qquad (1)$$

The LRNNs defined in this way will be referred to as *fully-connected*, as they contain rules for every *relation-entity-entity* triple. Obviously, when a high number of clusters is used, the number of rules of the form (1) may be prohibitively high. To address this, we can limit the triples for which such rules are added. In particular, we will consider LRNNs which restrict such rules to those of the following form:

$$w_{c_r c_{2i} c_{2i+1}} : Relation(R, A, B) \leftarrow IsA(R, c_r), IsA(A, c_{2i}), IsA(B, c_{2i+1}) \qquad (2)$$

where $c_1, c_2, \ldots, c_n$ are entity concepts. In fact the LRNNs with rules of this form can learn anything that can be learned by LRNNs with rules of the form (1) as long as they have enough rules.

In addition, to help the model learn symmetric and transitive relations (e.g. the "same-political-bloc" relation), we also add rules of the following form:

$$w_{c_r c_i' c_i'} : Relation(R, A, B) \leftarrow IsA(R, c_r), IsA(A, c_i'), IsA(B, c_i') \qquad (3)$$

Note that we do not need to explicitly consider these in the fully-connected model, as they are a special case of (1).

4   Prediction Using Learned Similarities

In this section we describe a LRNN model based on similarity degrees, for the same predictive task that was considered in the previous section. While the similarity degrees could be obtained from any source, we will use similarity degrees that have been obtained from the model described in the previous section, by taking advantage of the fact that the cluster membership degrees can be interpreted as defining a vector-space embedding. Rather than using the membership degrees directly, we will use the weights of the respective ground $IsA(e, c)$ rules, which, unlike the membership degrees, may also be negative[4]. In particular, the similarity degree between two entities is defined as the cosine similarity between the vector representation of these entities, with the coordinates of these vectors the soft memberships of the entity in each of the categories.

For each pair of entities (e_1, e_2) with similarity degree s, we add the following ground fact:

$$1.0 : Similar(e_1, e_2, s)$$

We furthermore add rules which encode a learnable transformation of the similarities into a score which is useful for the given predictive task:

$$w_{-1} : \qquad\qquad Similar(X, Y) \leftarrow Similar(X, Y, S), S \geq -1.0$$
$$w_{-0.9} : \qquad\qquad Similar(X, Y) \leftarrow Similar(X, Y, S), S \geq -0.9$$
$$\ldots$$
$$w_{0.9} : \qquad\qquad Similar(X, Y) \leftarrow Similar(X, Y, S), S \geq \;\;\; 0.9$$

Finally we add one rule of the following type for every relation r:

$$w_r : Relation(r, X, Y) \leftarrow Relation(r, V, W), Similar(X, V), Similar(Y, W).$$

Taking into account the aggregative nature of the used family of activation functions (cf. Sect. 2), these rules encode the intuition that in order to predict if X

---

[4] The membership degrees are simply obtained as applying sigmoids on the respective weights in this particular case, so the two representations essentially bear the same information.

and Y are in relation r, we could check how similar on average the entities known to be in this relation are to X and Y.

Naturally, not all relations can be accurately predicted by a model like the one described in this section. However, this similarity based approach is quite natural, and serves as an important illustrative example of how other strategies could be encoded (e.g. interpolation/extrapolation or reasoning by analogy).

5   Evaluation

5.1   Evaluation of the Model from Sect. 3.1

To evaluate the potential of the model proposed in Sect. 3.1, we have used the Animals dataset[5], which describes 50 animals in terms of 85 Boolean features, such as *fish, large, smelly, strong*, and *timid*. This dataset was originally created in [11], and was used among others for evaluating a related learning task in [7]. For both entities and properties, we have used two levels of categories, with in both cases three categories at the lowest level and two categories at the highest level.

Recall that we can view the category membership degrees as defining a vector-space embedding. Figures 1 and 2 show the first two principal components of this embedding for a number of entities and properties. We can see, for instance, that sea mammals are clustered together, and that predators tend to be separated from herbivores. In Fig. 2, we have highlighted two types of properties: colours and teeth types. Note that these do not form clusters (e.g. a cluster of colours) but they represent, as prototypes, different clusters of properties which tend to occur together. For instance, *blue* is surrounded by properties which typically hold for water mammals; white and red occur together with stripes, nocturnal, pads; gray occurs together with small and weak; etc. We also evaluated the predictive ability of this model. We randomly divided the facts from the dataset in two halves, trained the model on one half and tested it on the other one, obtaining AUC ROC of 0.77. We also performed an experiment with a 90–10 split, in order to be able to directly compare our results with those from [7]; we obtained the same AUC PR 0.8 as reported in [7] (and AUC ROC 0.86).

5.2   Evaluation of the Model from Sect. 3.2

In order to evaluate the relational method proposed in Sect. 3.2 we performed experiments with two relational datasets:[6] Nations and UMLS. These datasets have previously been used to evaluate statistical predicate invention methods in [7]. The Nations dataset contains a set of relations between pairs of nations and their features [15]. It consists of relations such as *ExportsTo* and *GivesEconomicAidTo*, as well as properties such as *Monarchy*. The dataset contains 14

---

[5] Downloaded from https://alchemy.cs.washington.edu/data/animals/.
[6] Downloaded from https://alchemy.cs.washington.edu/data/nations/ and from https://alchemy.cs.washington.edu/data/umls/.

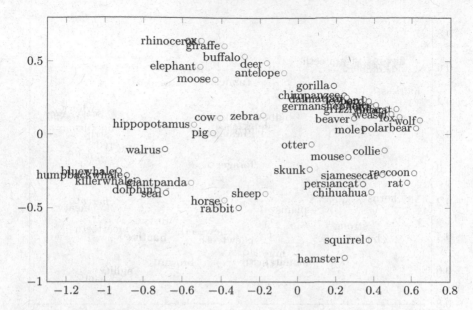

Fig. 1. Embedding of entities (animals, only a subset of entities is displayed). Several homogeneous groups of animals are highlighted: sea mammals (blue), large herbivores (green), rodents (violet), and other predators (red). (Color figure online)

nations, 56 relations and 111 properties. There are 2565 true ground atoms. The UMLS dataset contains data from the Unified Medical Language System, which is a biomedical ontology [8]. It contains 49 relations and 135 biomedical entities. There are 6529 true ground atoms in this dataset.

Initial experiments have revealed two trends. First, accuracy consistently improved when we increased the size of the LRNNs (contrarily to our expectation that overfitting might be a problem when increasing the size). Second, for a fixed number of entity, property and relation categories, adding the layer of more general concepts helps, but it also increased memory consumption and runtime. Therefore, in the experiments, we created LRNNs as large as possible which still fitted in memory. A consequence of this strategy is that the LRNNs with more than one layer of categories had fewer categories in total than their single-layer counterparts. Similar effects also took place for fully-connected LRNNs when compared to LRNNs with isolated rules of the form (2) and (3); therefore we did not consider fully connected LRNNs in our experiments.

For the Nations dataset, the largest single-layer LRNN which fitted in 40 GB of memory had 100 property categories, 100 entity categories and 50 relation categories. The cross-validated AUC ROC was 0.89 and AUC PR 0.74, which is within the standard error margin of the results obtained in [7]. The largest two-layer LRNN learned on this dataset had 20 property categories, 20 entity categories and 10 relation categories. Its cross-validated AUC ROC was 0.88 and AUC PR 0.7. For comparison, we also trained a single-layer LRNN with the

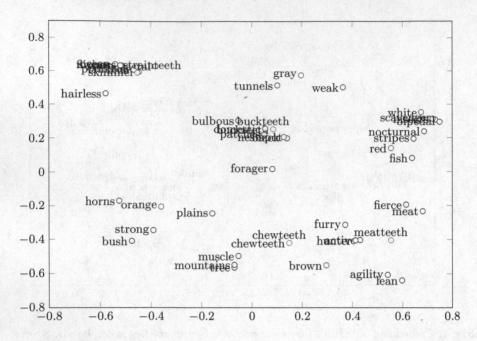

Fig. 2. Embedding of properties (only a subset of properties is displayed). Two representative groups of properties are shown in colour: colours (blue) and teeth-type (red). (Color figure online)

exact same number of each type of categories, which achieved AUC ROC 0.86 and AUC PR 0.67, which agrees with the above described general trends.

The first two principal components of the embeddings of the states are displayed in Fig. 3. When interpreting this embedding, note that this dataset relates to the political situation of 1950s.

For the UMLS dataset we used a LRNN with 100 entity categories and 50 relation categories. Due to the size of the dataset, consisting of a total of 893k ground facts and memory limitations, we only performed experiments with a largely subsampled training set, obtaining test AUC ROC 0.97 and AUC PR 0.76. This is a lower AUC PR than obtained by the method from [7], but it is close to the second-best method tested there and is better than the reported results for MLN structure learning.

5.3   Evaluation of the Model from Sect. 4

The evaluation of the model introduced in Sect. 4 primarily serves to estimate the usefulness of the embeddings learned by LRNNs. We have particulary focused on the embedding of countries, whose first two principal components are shown in Fig. 3. First, we have split the nations dataset [15] into equally large training and testing parts. We trained the relational model described in Sect. 3.2, extracted the learned cluster membership degrees as vector embeddings, and calculated

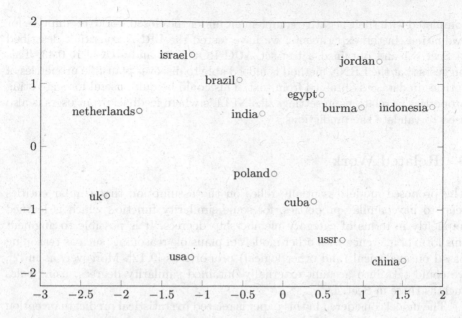

Fig. 3. Embeddings of countries as induced by learning from their geopolitical relations captured in the historical dataset [15]. A possible interpretation of the projection is displayed in colors, dividing them into communist (red), western (blue), and developing nations (green). (Color figure online)

their pairwise cosine similarities. We included these similarities as ground facts, together with all the true statements from the training part of the dataset. On top of these facts, we added the transformation and inference rules to form the model described in Sect. 4. We then trained this composite model on the same training part of the nations dataset that we used to obtain the embeddings, and evaluated its generalization ability on the remaining testing part. We obtained AUC ROC of 0.85 and 0.49 AUC PR, which is lower than the crossvalidated performance reported for the best models, but indirectly proves that the previously learned embeddings indeed carry useful information that may be subsequently reused for different predictive scenarios.

5.4   An Experiment with Real-Life Data from NELL

We have also evaluated the method on a real-life dataset. The main idea here was to analyse whether the LRNN models described in this paper could be used in an NLP pipeline to fill gaps in a knowledge base. To test this idea we downloaded a collection consisting of about 29k actors from NELL [10] with all their parental categories. For the experiments, we have subsampled the dataset to 2k actors. In the end, the number of different parental categories assigned to actors in this dataset turned out to be quite small. There were only 20 different categories such as *comedian* or *celebrity*, resulting into a dataset of 4k true ground facts, which we completed with their negative complement under the closed world assumption for evaluation. In the experiments, we have tested the LRNN construct described in Sect. 3.1 and obtained a test-set AUC ROC 0.84 and AUC PR 0.43. This suggests that the LRNN method is indeed able to discover plausible properties of entities in datasets obtained from text. This could be quite useful for suggesting properties or relations in settings like NELL's where feedback from users is also used to validate the predictions.

6   Related Work

The proposed model essentially relies on the assumption that similar entities tend to have similar properties, for some similarity function which is learned implicitly in terms of category membership degrees. It is possible to augment this form of inference with other models of plausible reasoning, such as reasoning based on analogical (and other logical) proportions [9,12]. Moreover, as in [2], we could take into account externally obtained similarity degrees, using rules such as those in Sect. 4.

The model considered in this paper is related to statistical predicate invention [7] which relies on jointly clustering entities and relations. The dual representation of entity and property categories is also reminiscent of formal concept analysis [5]. LRNNs themselves are also related to the long stream of research in neural-symbolic integration [1], previous work on using neural networks for relational learning [3], and more recent approaches such as [4,14].

7   Conclusions and Future Work

We have illustrated how the declarative and flexible nature of LRNNs can be used for easy encoding of non-trivial learning scenarios. The models that we considered in this paper jointly learn predictive categories of entities, their properties and relations between them. The main strength of this approach lies in the ease with which the model can be extended to more complicated settings, which is mainly due to the declarative nature of LRNNs. It seems remarkable that such a declarative approach is able to obtain results which are close to the state-of-the-art method from [7], without tailoring any part of the learning method to this particular problem setting.

Our main direction for future work will focus on making LRNNs more scalable, which, as indicated by the performed experiments, should also lead to improved predictive performance.

Acknowledgments. GS and FZ acknowledge support by project no. 17-26999S granted by the Czech Science Foundation. OK is supported by a grant from the Leverhulme Trust (RPG-2014-164). SS is supported by ERC Starting Grant 637277. Computational resources were provided by the CESNET LM2015042 and the CERIT Scientific Cloud LM2015085, provided under the programme "Projects of Large Research, Development, and Innovations Infrastructures".

References

1. Bader, S., Hitzler, P.: Dimensions of neural-symbolic integration-a structured survey. arXiv preprint cs/0511042 (2005)
2. Beltagy, I., Chau, C., Boleda, G., Garrette, D., Erk, K., Mooney, R.: Montague meets markov: deep semantics with probabilistic logical form. In: Proceedings of the *SEM, pp. 11–21 (2013)
3. Blockeel, H., Uwents, W.: Using neural networks for relational learning. In: ICML-2004 Workshop on Statistical Relational Learning and its Connection to Other Fields, pp. 23–28 (2004)
4. Cohen, W.W.: Tensorlog: a differentiable deductive database. arXiv preprint arXiv:1605.06523 (2016)
5. Ganter, B., Stumme, G., Wille, R. (eds.): Formal Concept Analysis: Foundations and Applications. LNCS (LNAI), vol. 3626. Springer, Heidelberg (2005)
6. Hájek, P.: Metamathematics of Fuzzy Logic, vol. 4. Springer, Dordrecht (1998)
7. Kok, S., Domingos, P.: Statistical predicate invention. In: Proceedings of the 24th International Conference on Machine Learning, pp. 433–440 (2007)
8. McCray, A.T.: An upper-level ontology for the biomedical domain. Comp. Funct. Genomics 4(1), 80–84 (2003)
9. Miclet, L., Bayoudh, S., Delhay, A.: Analogical dissimilarity: definition, algorithms and two experiments in machine learning. J. Artif. Intell. Res. 32, 793–824 (2008)
10. Mitchell, T.M., Cohen, W.W., Hruschka Jr., E.R., Talukdar, P.P., Betteridge, J., Carlson, A., Mishra, B.D., Gardner, M., Kisiel, B., Krishnamurthy, J., Lao, N., Mazaitis, K., Mohamed, T., Nakashole, N., Platanios, E.A., Ritter, A., Samadi, M., Settles, B., Wang, R.C., Wijaya, D.T., Gupta, A., Chen, X., Saparov, A., Greaves, M., Welling, J.: Never-ending learning. In: Proceedings of the Twenty-Ninth AAAI Conference on Artificial Intelligence, January 25–30, 2015, Austin, Texas, USA, pp. 2302–2310 (2015)
11. Daniel, N.O., Stern, J., Wilkie, O., Stob, M., Smith, E.E.: Default probability. Cogn. Sci. 15(2), 251–269 (1991)
12. Prade, H., Richard, G.: Reasoning with logical proportions. In: Twelfth International Conference on the Principles of Knowledge Representation and Reasoning (2010)
13. Richardson, M., Domingos, P.: Markov logic networks. Mach. Learn. 62(1–2), 107–136 (2006)
14. Rocktäschel, T., Riedel, S.: Learning knowledge base inference with neural theorem provers. In: NAACL Workshop on Automated Knowledge Base Construction (AKBC) (2016)
15. Rummel, R.J.: The dimensionality of nations project: attributes of nations and behavior of nations dyads, pp. 1950–1965. Number 5409. Inter-University Consortium for Political Research (1976)
16. Šourek, G., Aschenbrenner, V., Železný, F., Kuželka, O.: Lifted relational neural networks. In: Proceedings of the NIPS Workshop on Cognitive Computation: Integrating Neural and Symbolic Approaches (2015)

Generation of Near-Optimal Solutions Using ILP-Guided Sampling

Ashwin Srinivasan[1], Gautam Shroff[2], Lovekesh Vig[2],
and Sarmimala Saikia[2(✉)]

[1] Department of Computer Science and Information Systems BITS Pilani,
Goa Campus, Goa 403726, India
ashwin@goa.bits-pilani.ac.in
[2] TCS Research, New Delhi, India
{gautam.shroff,lovekesh.vig,sarmimala.saikia}@tcs.com

Abstract. Our interest in this paper is in optimisation problems that
are intractable to solve by direct numerical optimisation, but nevertheless have significant amounts of relevant domain-specific knowledge. The
category of heuristic search techniques known as estimation of distribution algorithms (EDAs) seek to incrementally sample from probability
distributions in which optimal (or near-optimal) solutions have increas-
ingly higher probabilities. Can we use domain knowledge to assist the
estimation of these distributions? To answer this in the affirmative, we
need: (a) a general-purpose technique for the incorporation of domain
knowledge when constructing models for optimal values; and (b) a way
of using these models to generate new data samples. Here we investigate a combination of the use of Inductive Logic Programming (ILP) for
(a), and standard logic-programming machinery to generate new samples for (b). Specifically, on each iteration of distribution estimation, an
ILP engine is used to construct a model for good solutions. The resulting theory is then used to guide the generation of new data instances,
which are now restricted to those derivable using the ILP model in conjunction with the background knowledge). We demonstrate the approach
on two optimisation problems (predicting optimal depth-of-win for the
KRK endgame, and job-shop scheduling). Our results are promising: (a)
On each iteration of distribution estimation, samples obtained with an
ILP theory have a substantially greater proportion of good solutions
than samples without a theory; and (b) On termination of distribution
estimation, samples obtained with an ILP theory contain more near-
optimal samples than samples without a theory. Taken together, these
results suggest that the use of ILP-constructed theories could be a useful
technique for incorporating complex domain-knowledge into estimation
distribution procedures.

Keywords: Domain-knowledge guided optimisation · Estimation of
distribution · Inductive logic programming

© Springer International Publishing AG 2017
J. Cussens and A. Russo (Eds.): ILP 2016, LNAI 10326, pp. 120–131, 2017.
DOI: 10.1007/978-3-319-63342-8_10

1   Introduction

There are many real-world planning problems for which domain knowledge is qualitative, and not easily encoded in a form suitable for numerical optimisation. Here, for instance, are some guiding principles that are followed by the Australian Rail Track Corporation when scheduling trains: (1) If a "healthy" train is running late, it should be given equal preference to other healthy trains; (2) A higher priority train should be given preference to a lower priority train, provided the delay to the lower priority train is kept to a minimum; and so on. It is evident from this that train-scheduling may benefit from knowing if a train is "healthy", what a train's priority is, and so on. But are priorities and train-health fixed, irrespective of the context? What values constitute acceptable delays to a low-priority train? Generating good train-schedules will require a combination of quantitative knowledge of a train's running times and qualitative knowledge about the train in isolation, and in relation to other trains. In this paper, we propose a heuristic search method, that comes under the broad category of an estimation distribution algorithm (EDA). EDAs iteratively generates better solutions to the optimisation problem using machine-constructed models. Usually EDA's have used generative probabilistic models, such as Bayesian Networks, where domain-knowledge needs to be translated into prior distributions and/or network topology. In this paper, we are concerned with problems for which such a translation is not evident. Our interest in ILP is that it presents perhaps one of the most flexible ways to use domain-knowledge when constructing models.

What can this form of optimisation do differently? First, there is the straightforward difference to standard optimisation, arising from the use of domain-knowledge in first-order logic. Traditionally, optimisation methods have required domain knowledge to be in the form of linear inequalities. This quickly becomes complicated. For example, $y = x_1 \oplus x_2$ requires the inequalities $y \leq x_1 + x_2 \wedge y \leq 2 - x_1 - x_2 \wedge y \geq x_1 - x_2 \wedge y \geq x_2 - x_1 \wedge y \geq 0 \wedge y \leq 1$. As a statement in logic, the relation is clearly trivial: so, we would expect to do better on problems for which domain-knowledge is far easier to express in logical form than as linear constraints (of course, one could consider non-linear constraints, but then the optimisation problem becomes much harder). Secondly, there is the difference arising from constructing models in first-order logic. Most probabilistic models used in EDA only allow to use models that involve statements about propositions. This restricts the expressivity of the models, or requires large numbers of propositions representing pre-defined relations. We would therefore expect to do better on problems that require models that involve relationships amongst background predicates that are not easy to know beforehand.

The rest of the paper is organised as follows. Section 2 provides a brief description of the EDA method we use for optimisation problems. Section 2.1 describes how ILP can be used within the iterative loop of an EDA, including a procedure for sampling data instances entailed by the ILP-theory. Section 3 describes an empirical evaluation followed by conclusions in Sect. 4.

2   EDA for Optimisation

The basic EDA approach we use is the one proposed by the MIMIC algorithm
[5]. Assuming that we are looking to minimise an objective function $F(\mathbf{x})$, where
$\mathbf{x}$ is an instance from some instance-space $\mathcal{X}$, the approach first constructs an
appropriate machine-learning model to discriminate between samples of lower
and higher value, i.e., $F(\mathbf{x}) \leq \theta$ and $F(\mathbf{x}) > \theta$, and then sampling from this
model to generate a population for the next iteration, while also lowering θ.
This is described by the procedure in Fig. 1.

Procedure EOMS: Evolutionary Optimisation using Model-Assisted Sampling
1. Initialize population $P := \{\mathbf{x}_i\}$; $\theta := \theta_0$
2. while not converged do
   (a) for all $\mathbf{x}_i$ in P $label(\mathbf{x}_i) := 1$ if $F(\mathbf{x}_i) \leq \theta$ else $label(\mathbf{x}_i) := 0$
   (b) train model M to discriminate between 1 and 0 labels i.e., $P(\mathbf{x} : label(\mathbf{x}) = 1|M) > P(\mathbf{x} : label(\mathbf{x}) = 0|M)$
   (c) regenerate P by repeated sampling using model M
   (d) reduce threshold θ
3. return P

Fig. 1. Evolutionary optimisation using machine-learning models to guide sampling.

2.1   ILP-assisted Evolutionary Optimisation

We propose to use ILP as the model construction technique in the EOMS procedure, since it provides an extremely flexible way to construct models using
domain-knowledge. On the face of it, this would seem to pose a difficulty for the
sampling step: how are we to generate new instances that are entailed by an ILP-
constructed model? There are two straightforward options. First, if we have an
enumerator of the instance space $\mathcal{X}$, then we could resort to a form of rejection-
sampling. Second, we can restrict ILP-theories for any predicate to generative
clauses, which allows the theories to be used generatively.[1], using standard logic-
programming inference machinery to generate instances of the success-set of each
predicate. Instances obtained in this manner are selected with some probability
to achieve a non-uniform sampling of the success-set. Both these methods are
viable for the purposes of this paper, but in general, we expect that more sophisticated ways of sampling would be needed: see for example [4]. The procedure
EOIS in Fig. 2 is a refinement of the EOMS procedure above.

In EOIS, $ilp(B, E^+, E^-)$ is an ILP algorithm that returns a theory M s.t.
$B \wedge M \models E^+$; $B \wedge M$ is inconsistent with the E^- only to the extent allowed
by constraints in B; and $sample(n, M, B)$ returns a set of at most n instances
entailed by $B \wedge M$, if $M \neq \emptyset$. If $M = \emptyset$, it returns a random selection of
n instances from the instance-space. In general, we require $sample$ to draw

---

[1] A syntactic way to do this is by adding constraints to the body of the clause that
impose range (that is, type) restrictions on the variables in the head: see [6].

Procedure EOIS: Evolutionary Optimisation using ILP-Assisted Sampling

Given: (a) Background knowledge B; (b) an upper-bound θ^* on the cost of acceptable solutions; (c) a decreasing sequence of cost-values $\theta_1, \theta_2, \ldots, \theta_n$ s.t. $\theta_1 \geq \theta^* \geq \theta_n$; and (d) an upper-bound on the sample size n

1. Let $M_0 := \emptyset$ and $P_0 := sample(n, M_0, B)$
2. Let $k = 1$
3. while $(\theta_k \geq \theta^*)$ do
    (a) $E_k^+ := \{\mathbf{x}_i : \mathbf{x}_i \in P$ and $F(\mathbf{x}_i) \leq \theta_k\}$ and $E_k^- := \{\mathbf{x}_i : \mathbf{x}_i \in P$ and $F(\mathbf{x}_i) > \theta_k\}$
    (b) $M_k := ilp(B, E_k^+, E_k^-)$
    (c) $P_k := sample(n, M_k, B)$
    (d) increment k
4. return P_{k-1}

Fig. 2. Evolutionary optimisation using ILP models to guide sampling.

instances from the success-set of the ILP-constructed theory, since these are the "good" solutions entailed by the model on each iteration (see [4] for techniques for doing this). Here, we make it by providing an initial sample to EOIS as input when $M = \emptyset$ (to prevent biasing future iterations: this sample is obtained by uniform random selection from the instance space). For subsequent steps, we assume the availability of a generator that returns a sample of the success-set of target-predicate. That is, when $M \neq \emptyset$, $sample$ returns the set $S = \{e_i : 1 \leq i \leq n \ e_i \in \mathcal{X}$ and $B \wedge M \vdash e$ and $Pr(e_i) \geq \delta\}$, where δ is some probability threshold (correctly therefore $sample(n, M, B)$ should be $sample(n, \delta, M, B)$). Thus if $\delta = 1$, the first n instances derived (or fewer if there are less) using SLD-resolution with $B \wedge M$ will be selected. Finally, we note that on iterations $k \geq 1$, we can use the data $P_0 \cup \cdots \cup P_{k-1}$ to obtain training examples E_k^+ and E_k^- since the actual costs for the P's have already been computed. For clarity, this detail has been omitted.

3 Empirical Evaluation

3.1 Aims

Our aims in the empirical evaluation are to investigate the following conjectures:

(1) On each iteration, the EOIS procedure will yield better samples than simple random sampling of the instance-space; and
(2) On termination, the EOIS procedure will yield more near-optimal instances than simple random sampling of the same number of instances as used for constructing the model.

It is relevant here to clarify what the comparisons are intended in the statements above. Conjecture (1) is essentially a statement about the gain in precision obtained by using the model. Let us denote $Pr(F(\mathbf{x}) \leq \theta)$ the probability of generating an instance $\mathbf{x}$ with cost at most θ without a model to guide sampling (that is, using simple random sampling of the instance space), and by $Pr(F(\mathbf{x}) \leq \theta | M_{k,B})$ the probability of obtaining such an instance with an ILP-constructed model $M_{k,B}$ obtained on iteration k of the EOIS procedure using some domain-knowledge B. (note if $M_{k,B} = \emptyset$, then we will mean $Pr(F(\mathbf{x}) \leq \theta | M_{k,B}) = Pr(F(\mathbf{x}) \leq \theta)$). Then for (1) to hold, we would require $Pr(F(\mathbf{x}) \leq \theta_k | M_{k,B}) > Pr(F(\mathbf{x}) \leq \theta_k)$, given some relevant B. We will estimate the probability on the lhs from the sample generated using the model, and the probability on the rhs from the datasets provided.

Conjecture (2) is related to the gain in recall obtained by using the model, although it is more practical to examine actual numbers of near-optimal instances (true-positives in the usual terminology). We will compare the numbers of near-optimal in the sample generated by the model to those obtained using random sampling.

3.2   Materials

Data. We use two synthetic datasets, one arising from the KRK chess endgame (an endgame with just White King, White Rook and Black King on the board), and the other a restricted, but nevertheless hard 5×5 job-shop scheduling (scheduling 5 jobs taking varying lengths of time onto 5 machines, each capable of processing just one task at a time).

The optimisation problem we examine for the KRK endgame is to predict the depth-of-win with optimal play [1]. Although aspect of the endgame has not been as popular in ILP as task of predicting "White-to-move position is illegal" [2,8], it offers a number of advantages as a *Drosophila* for optimisation problems of the kind we are interested. First, as with other chess endgames, KRK-win is a complex, enumerable domain for which there is complete, noise-free data. Second, optimal "costs" are known for all data instances. Third, the problem has been studied by chess-experts at least since Torres y Quevado built a machine, in 1910, capable of playing the KRK endgame. This has resulted in a substantial amount of domain-specific knowledge. We direct the reader to [3] for the history of automated methods for the KRK-endgame. For us, it suffices to treat the problem as a form of optimisation, with the cost being the depth-of-win with Black-to-move, assuming minimax-optimal play. In principle, there are $64^3 \approx 260,000$ possible positions for the KRK endgame, not all legal. Removing illegal positions, and redundancies arising from symmetries of the board reduces the size of the instance space to about $28,000$ and the distribution shown in Fig. 3(a). The sampling task here is to generate instances with depth-of-win equal to 0. Simple random sampling has a probability of about $1/1000$ of generating such an instance once redundancies are removed.

The job-shop scheduling problem is less controlled than the chess endgame, but is nevertheless representative of many real-life applications (like scheduling trains), which are, in general, known to be computationally hard. We use a job-shop problem with five jobs, each consisting of five tasks that need to be executed in order. These 25 tasks are to be performed using 5 machines, each capable of

Cost	Instances	Cost	Instances
0	27 (0.001)	9	1712 (0.196)
1	78 (0.004)	10	1985 (0.267)
2	246 (0.012)	11	2854 (0.368)
3	81 (0.152)	12	3597 (0.497)
4	198 (0.022)	13	4194 (0.646)
5	471 (0.039)	14	4553 (0.808)
6	592 (0.060)	15	2166 (0.886)
7	683 (0.084)	16	390 (0.899)
8	1433 (0.136)	draw	2796 (1.0)

Total Instances: 28056

(a) Chess

Cost	Instances	Cost	Instances
400–500	10 (0.0001)	1000–1100	24067 (0.748)
500–600	294 (0.003)	1100–1200	15913 (0.907)
600–700	2186 (0.025)	1200–1300	7025 (0.978)
700–800	7744 (0.102)	1300–1400	1818 (0.996)
800–900	16398 (0.266)	1400–1500	345 (0.999)
900–1000	24135 (0.508)	1500–1700	66 (1.0)

Total Instances: 100000

(b) Job-Shop

Fig. 3. Distribution of cost values. Numbers in parentheses are cumulative frequencies.

performing a particular task, albeit for any of the jobs. A 5×5 matrix defines how long task j of job i takes to execute on machine j.

Data instances for Chess are in the form of 6-tuples, representing the rank and file (X and Y values) of the 3 pieces involved. At each iteration k of the EOIS procedure, some instances with depth-of-win $\leq \theta_k$ and the rest with depth-of-win $> \theta_k$ are used to construct a model.[2]

Data instances for Job-Shop are in the form of schedules defining the sequence in which tasks of different jobs are performed on each machine, along with the total cost (i.e., time duration) implied by the schedule. On iteration i of the EOIS procedure, models are to be constructed to predict if the cost of schedule will be $\leq \theta_i$ or otherwise.[3]

Background Knowledge. For Chess, background predicates encode the following (WK denotes the White King, WR the White Rook, and BK the Black King): (a) Distance between pieces WK-BK, WK-BK, WK-WR; (b) File and distance patterns: WR-BK, WK-WR, WK-BK; (c) "Alignment distance": WR-BK; (d) Adjacency patterns: WK-WR, WK-BK, WR-BK; (e) "Between" patterns: WR between WK and BK, WK between WR and BK, BK between WK and WR; (f) Distance to closest edge: BK; (g) Distance to closest corner: BK; (h) Distance to centre: WK; and (i) Inter-piece patterns: Kings in opposition, Kings almost-in-opposition, L-shaped pattern. We direct the reader to [3] for the history of using these concepts, and their definitions.

For Job-Shop, background predicates encode: (a) schedule job J "early" on machine M (early means first or second); (b) schedule job J "late" on machine

---

[2] The θ_k values are pre-computed assuming optimum play. We note that when constructing a model on iteration k, it is permissible to use all instances used on iterations $1, 2, \ldots, (k-1)$ to obtain data for model-construction.

[3] The total cost of a schedule includes any idle-time, since for each job, a task before the next one can be started for that job. Again, on iteration i, it is permissible to use data from previous iterations.

M (late means last or second-last); (c) job J has the fastest task for machine M; (d) job J has the slowest task for machine M; (e) job J has a fast task for machine M (fast means the fastest or second-fastest); (f) Job J has a slow task for machine M (slow means slowest or second-slowest); (g) Waiting time for machine M; (h) Total waiting time; (i) Time taken before executing a task on a machine. Correctly, the predicates for (g)–(i) encode upper and lower bounds on times, using the standard inequality predicates $\leq$ and $\geq$.

Algorithms and Machines. The ILP-engine we use is Aleph (Version 6, available from A.S. on request). All ILP theories were constructed on an Intel Core i7 laptop computer, using VMware virtual machine running Fedora 13, with an allocation of 2GB for the virtual machine. The Prolog compiler used was Yap, version 6.1.3[4].

3.3   Method

Our method is straightforward:

> For each optimisation problem, and domain-knowledge B:
> Using a sequence of threshold values $\langle \theta_1, \theta_2, \ldots, \theta_n \rangle$ on iteration k ($1 \leq k \leq n$) for the EOIS procedure:
> 1. Obtain an estimate of $Pr(F(\mathbf{x}) \leq \theta_k)$ using a simple random sample from the instance space;
> 2. Obtain an estimate of $Pr(F(\mathbf{x}) \leq \theta_k | M_{k,B})$ by constructing an ILP model for discriminating between $F(\mathbf{x}) \leq \theta_k$ and $F(\mathbf{x}) > \theta_k$
> 3. Compute the ratio of $Pr(F(\mathbf{x}) \leq \theta_k | M_{k,B})$ to $P(F(\mathbf{x}) \leq \theta_k)$

The following details are relevant:

- The sequence of thresholds for Chess are $\langle 8, 4, 0 \rangle$. For Job-Shop, this sequence is $\langle 1000, 750, 600 \rangle$; Thus, $\theta^* = 0$ for Chess and 600 for Job-Shop, which means we require exactly optimal solutions for Chess.
- Experience with the use of ILP engine used here (Aleph) suggests that the most sensitive parameter is the one defining a lower-bound on the precision of acceptable clauses (the *minacc* setting in Aleph). We report experimental results obtained with *minacc* $= 0.7$, which has been used in previous experiments with the KRK dataset. The background knowledge for Job-Shop does not appear to be sufficiently powerful to allow the identification of good theories with short clauses. That is, the usual Aleph setting of upto 4 literals per clause leaves most of the training data ungeneralised. We therefore allow an upper-bound of upto 10 literals for Job-Shop, with a corresponding increase in the number of search nodes to 10000 (Chess uses the default setting of 4 and 5000 for these parameters).

---

[4] http://www.dcc.fc.up.pt/~vsc/Yap/.

– In the EOIS procedure, the bound on sample size n is 1000. The initial sample is obtained using a uniform distribution over all instances. Let us call this P_0. On the first iteration of EOIS ($k = 1$), the datasets $E_1{}^+$ and $E_1{}^-$ are obtained by computing the (actual) costs for instances in P_0, and an ILP model $M_{1,B}$, or simply M_1, constructed. To obtain a sample of instances entailed by the model $M_{k,B}$ we use the *sample* function with a δ value of 1.0. That is, the first n unique instances (or fewer, if less) obtained by employing SLD-resolution on $B \wedge M_{k,B}$ are taken as the sample P_k. For Chess, it has been possible to ensure that the logic-programs involved are generative. Thus, we are able to use $M_{k,B}$ directly as a generator of instances entailed by the $B \wedge M_{k,B}$. For Job-Shop, we employ rejection-sampling instead. That is, we randomly draw from the instance-space, and then check to see if it is entailed by $B \wedge M_{k,B}$. Both approaches have not proved to be especially inefficient, probably because the instance-spaces are small. On each iteration k, an estimate of $Pr(F(\mathbf{x}) \leq \theta_k)$ can be obtained from the empirical frequency distribution of instances with values $\leq \theta_k$ and $> \theta_k$. For the synthetic problems here, these estimates are in Fig. 3. For $Pr(F(\mathbf{x}) \leq \theta_k | M_{k,B})$, we use obtain the frequency of $F(\mathbf{x}) \leq \theta_k$ in P_k

– Readers will recognise that the ratio of $Pr(F(\mathbf{x}) \leq \theta_k | M_{k,B})$ to $P(F(\mathbf{x}) \leq \theta_k)$ is equivalent to computing the gain in precision obtained by using an ILP model over random selection. Specifically, if this ratio is approximately 1, then there is no value in using the ILP model. The probabilities computed also provide one way of estimating sampling efficiency of the models (the higher the probability, the fewer samples will be needed to obtain an instance $\mathbf{x}$ with $F(\mathbf{x}) \leq \theta_k$).

3.4   Results

Results relevant to conjectures (1) and (2) are tabulated in Fig. 4 and Fig. 5. The principal conclusions that can drawn from the results are these:

(1) For both problems, and every threshold value θ_k, the probabilty of obtaining instances with cost at most θ_k with model-guided sampling is substantially higher than without a model. This provides evidence that model-guided sampling results in better samples than simple random sampling (Conjecture 1);

(2) For both problems and every threshold value θ_k, samples obtained with model-guided sampling contain a substantially higher number of near-optimal instances than samples obtained without a model (Conjecture 2).

We note also that all results have been obtained by sampling a small portion of the instance space (about 10 % for Chess, and about 3 % for Job-Shop).

We now examine the result in more detail. It is evident that the performance on the Job-Shop domain is not as good as on Chess. The natural question that arises is: Why is this so? We conjecture that this is a consequence of the background knowledge for Job-Shop not being as relevant to low cost values, as was

Model	$Pr(F(\mathbf{x}) \leq \theta_k \vert M_k)$		
	$k = 1$	$k = 2$	$k = 3$
None	0.136	0.022	0.001
ILP	0.816	0.462	0.409
	(6.0)	(21.0)	(409.0)

(a) Chess

Model	$Pr(F(\mathbf{x}) \leq \theta_k \vert M_k)$		
	$k = 1$	$k = 2$	$k = 3$
None	0.507	0.025	0.003
ILP	0.647	0.171	0.080
	(1.3)	(6.8)	(26.7)

(b) Job-Shop

Fig. 4. Probabilities of obtaining good instances $\mathbf{x}$ for each iteration k of the EOIS procedure. That is, the column $k = 1$ denotes $P(F(\mathbf{x}) \leq \theta_1$ after iteration 1; the column $k = 2$ denotes $P(F(\mathbf{x}) \leq \theta_2$ after iteration 2 and so on. In effect, this is an estimate of the precision when predicting $F(\mathbf{x}) \leq \theta_k$. "None" in the model column stands for probabilities of the instances, corresponding to simple random sampling ($M_k = \emptyset$). The number in parentheses below each ILP entry denotes the ratio of that entry against the corresponding entry for "None". This represents the gain in precision of using the ILP model over simple random sampling.

Model	Near-Optimal Instances		
	$k = 1$	$k = 2$	$k = 3$
None	1/27	2/27	3/27
ILP	11/27	22/27	27/27
	(1000)	(1964)	(2549)

(a) Chess

Model	Near-Optimal Instances		
	$k = 1$	$k = 2$	$k = 3$
None	3/304	6/304	9/304
ILP	6/304	28/304	36/304
	(1000)	(1987)	(2895)

(b) Job-Shop

Fig. 5. Fraction of near-optimal instances ($F(\mathbf{x}) \leq \theta^*$) generated on each iteration of EOIS. In effect, this is an estimate of the recall (true-positive rate, or sensitivity) when predicting $F(\mathbf{x}) \leq \theta^*$. The fraction a/b denotes that a instances of b are generated. The numbers in parentheses denote the number of training instances used by the ILP engine. The values with "None" are the numbers expected by sampling the same number of training instances used for training the ILP engine.

the case for Chess. Some evidence for this was already apparent when we had to increase the lengths of clauses allowed for the ILP engine (this is usually a sign that the background knowledge is somewhat low-level). In contrast, with Chess, some of the concepts refer specifically to "cornering" the Black King, with a view of ending the game as soon as possible. We would expect these predicates to be especially useful for positions at depths-of-win near 0. Evidence of the unreliable performance of the EOIS procedure in Chess, with irrelevant background knowledge is in Fig. 6. These results suggest a refinement to the conclusions we can draw from the use of EOIS, namely: we expect the EOIS procedure to be less effective if the background knowledge is not very relevant to low-cost solutions.

Finally, we note that the experiments with synthetic data have ignored an important aspect of the optimisation problem, namely the time taken to obtain

Background	$Pr(F(\mathbf{x}) \leq \theta_k \| M_{k,B})$		
B	$k = 1$	$k = 2$	$k = 3$
B_{low}	0.658	0.417	0.0
B_{high}	0.816	0.462	0.409

(a)

Background	Near-Optimal Instances		
B	$k = 1$	$k = 2$	$k = 3$
B_{low}	17/27	4/27	0/27
	(1000)	(1959)	(2581)
B_{high}	11/27	17/27	27/27
	(1000)	(1964)	(2549)

(b)

Fig. 6. Precision (a), and recall of near-optimal instances (b) of the EOIS procedure with background knowledge of low relevance to near-optimal solutions. The results are for Chess, with B_{low} denoting background predicates that simply define the geometry of the board, using the predicates *less_than* and *adjacent*. These predicates form the background knowledge for most ILP applications to the problem of detecting illegal positions in the KRK endgame. B_{high} denotes background used previously, with high relevance to low depths-of-win. The numbers in parentheses in (b) are the number of training instances as before.

the value of the objective function for each data instance. Clearly, there is a trade-off between the time taken to construct a model, and the time taken to simply draw instances without a model. To address this trade-off we would have to estimate the number of instances that need to be randomly sampled to obtain the same numbers of near-optimal instances as with model-assisted EDA; and to compare: (a) the time to obtain values of the objective function for randomly-sampled instances; and (b) the time taken to obtain the corresponding values for the training data used to construct models and the total time taken for model construction. For model-construction to be beneficial, clearly the time in (b) has to be less than (a). For the problems here, random sampling require approximately 4 (JobShop) to 10 (Chess) times as many samples to obtain the same numbers of near-optimal instances. The times for theory-construction are small enough to expect that (b) is less than (a) for these ratios.

4   Concluding Remarks

It is uncommon to see ILP applied to optimisation problems. The use of ILP-constructed theories within an evolutionary optimisation procedure is one answer to the question of how to use ILP for optimisation (but not the only answer: recent work in [7], for example, suggests using ILP-constructed clauses as soft-constraints to a constraint solver). This requires the ILP model to be used generatively, which is not common practice in ILP. In this paper we have been able to do this by a combination of careful definition of the background knowledge, and adding range-restrictions to clauses constructed by ILP. Our results provide evidence that this combination of ILP and logic-programming provides one way of incorporating complex, but relevant domain-knowledge into evolutionary optimisation.

Concerning the specific problems examined here, it is possible that we could have directly constructed a model discriminating near-optimal instances from the rest using ILP alone. The focus of this paper however is on a different question, namely, whether evolutionary optimisation methods can benefit from the use of ILP. The results here should therefore be seen as evidence of improvements possible in an EDA technique when it includes ILP-assisted models. In turn, this evidence could be of relevance for problems where ILP models alone would be insufficient, and we would have to resort to sampling-based methods.

There are several ways in which the work here could be extended. The most immediate is to examine ways of sampling by using techniques developed in probabilistic ILP. Indeed, the principal conjecture of the paper is that the use of models constructed by any form of learning that allows the inclusion domain-knowledge can greatly improve the sampling efficiency of EDA methods. We have provided evidence for this conjecture using a classical ILP method. Given these results, we can expect first-order learning that is capable of using domain-knowledge and constructing rules that can allow a non-uniform sampling of ground instances (for example, through the incorporation of probabilities with the rules) will provide even better results. We would recommend this as the next step in this line of work.

It is of interest also to consider whether there are any gains to be made by re-use of theories (currently, we re-use data, but re-learn theories from scratch on each new iteration of the EOIS procedure). There is the straightforward approach to this, of simply providing theories constructed earlier as background knowledge for subsequent iterations. A more ambitious variant would retain some portions of the earlier theory (those clauses that entail the current set of positive instances and none, or only few, of the negative instances, for example), thus reducing the model-construction effort.

Acknowledgments. A.S. is a Visiting Professor at the Department of Computer Science, University of Oxford; and Visiting Professorial Fellow at the School of CSE, UNSW.

References

1. Bain, M., Muggleton, S.: Learning optimal chess strategies. In: Furukawa, K., Michie, D., Muggleton, S. (eds.) Machine Intelligence 13, pp. 291–309. Oxford University Press Inc, New York (1995)
2. Bain, M.: Learning logical exceptions in chess. Ph.D. Thesis, University of Strathclyde (1994)
3. Breda, G.: KRK Chess Endgame Database Knowledge Extraction and Compression. Diploma Thesis, Technische Universität, Darmstadt (2006)
4. Cussens, J.: Stochastic logic programs: sampling, inference and applications. In: Proceedings of the Sixteenth Conference on Uncertainty in Artificial Intelligence, UAI 2000, pp. 115–122, San Francisco, CA, USA. Morgan Kaufmann Publishers Inc (2000)

5. De Bonet, J.S., Isbell, C.L., Viola, P. et al.: Mimic: finding optima by estimating probability densities. In: Advances in Neural Information Processing Systems, pp. 424–430 (1997)
6. Muggleton, S., Feng, C.: Efficient induction of logic programs. In: Muggleton, S. (ed.) Inductive Logic Programming, pp. 281–298. Academic Press, London (1992)
7. Shiina, Y., Ohwada, H.: Using machine-generated soft constraints for roster problems. In: Muggleton, S., Watanabe, H. (eds.) Latest Advances in Inductive Logic Programming, pp. 227–234. World Scientific Press (2014)
8. Srinivasan, A., Muggleton, S.H., Bain, M.E.: Distinguishing noise from exceptions in non-monotonic learning. In: Muggleton, S., Furukawa, K. (eds.) Second International Inductive Logic Programming Workshop (ILP92). Institute for New Generation Computer Technology (1992)

Author Index

Artikis, Alexander 27

Besold, Tarek 52
Blask, Erik 14

Côrte-Real, Joana 1

Dutra, Inês 1

Gad-Elrab, Mohamed H. 94

Kersting, Kristian 40
Khot, Tushar 14
Kumaraswamy, Raksha 40
Kuželka, Ondřej 108

Lisi, Francesca A. 94

Malec, Marcin 14
Manandhar, Suresh 108
Matsumoto, Satoshi 68
Michelioudakis, Evangelos 27
Muggleton, Stephen 52

Nagy, James 14
Natarajan, Sriraam 14, 40, 81

Odom, Phillip 40

Paliouras, Georgios 27

Rocha, Ricardo 1

Saikia, Sarmimala 120
Schmid, Ute 52
Schockaert, Steven 108
Shavlik, Jude 81
Shoudai, Takayoshi 68
Shroff, Gautam 120
Soni, Ameet 81
Šourek, Gustav 108
Srinivasan, Ashwin 120
Stepanova, Daria 94
Suzuki, Yusuke 68

Tamaddoni-Nezhad, Alireza 52
Tran, Hai Dang 94

Vig, Lovekesh 120
Viswanathan, Dileep 81

Weikum, Gerhard 94

Železný, Filip 108
Zeller, Christina 52

Printed in the United States
By Bookmasters